HISTORY OF
**SCIENCE · AND ·
TECHNOLOGY**
REPRINT SERIES

A History of Cytology

Arthur Hughes

IOWA STATE UNIVERSITY PRESS ▪ Ames

Arthur Hughes became a Sir Halley Stewart Research Fellow at the Strangeways Research Laboratory after graduation from Cambridge University. He then became a University Lecturer in anatomy at Cambridge University.

This edition published in 1989 by Iowa State University Press, Ames, Iowa 50010

Text reprinted from the original without correction by arrangement with Harper & Row

Printed in the United States of America

∞ This book is printed on acid-free paper.

History of Science and Technology Reprint Series

First printing, 1989

Library of Congress Cataloging-in-Publication Data

Hughes, Arthur Frederick William, 1908–
A history of cytology/Arthur Hughes.
p. cm.—(History of science and technology reprint series)
Reprint. Originally published: London, New York: Abelard-Schuman, 1959.
Includes bibliographical references.
ISBN 0–8138–0858–8
1. Cytology—History. I. Title. II. Series.
QH577.H8 1989
574.87′09—dc20 89–24419

for Marjorie

Contents

Preface

An account of so small a fragment of the sum total of human experience as the development of microscopical research can hardly hope to gratify the general historian to any appreciable extent. It is of no special significance, for instance, that the discovery of Brownian motion preceded the emancipation of Roman Catholics in England by a single year. It is certainly true that social history now recognises the importance of scientific discovery as a factor in economic progress, but in this tends to concentrate on such knowledge 'as hath a tendency for use', as Robert Boyle put it. Thus pure research is overshadowed by technology in the story of past centuries as in the expanding universities of the present day.

The history of any branch of learning, however specialised or arcane, has nevertheless its own purpose. The invention of the microscope in the seventeenth century disclosed a new world, that of the hitherto invisibly small, and summoned the student of nature to explore it. Efforts to meet this challenge have continued ever since; though always the horizon of final comprehension steadily recedes with each new forward step. Robert Hooke, with his imperfect compound microscope, baffled in the search for Nature's 'appropriated instruments and contrivances to bring her designs and ends to pass', stands here beside the contemporary electron microscopist still without sight of the physical gene. If those who pursue these new and exciting methods of research were more aware of their station in a historical sequence they might perhaps be less inclined to deck their findings in new and unnecessary jargon. Robert Hooke was content with the noble English of his day; but we now read of 'profiles' and of 'fine structure appearances' without the conviction that a fresh gain in magnification necessitates a concomitant grandiloquence.

It is abundantly clear that the motions of the human mind are similar whatever optical means lie before it. During the few centuries of microscopical endeavour similar ideas have often recurred, though sometimes in ignorance of their previous appearances. Martin Barry in the early 1840's regarded the nucleolus as a centre of synthetic activity in the cell rather over a century before T. O. Caspersson made a similar claim. Sometimes the same terms, such as 'microsome' or 'ergastoplasm', are re-introduced, though with reference to a new order of size. A technique may be revived after a long interval of disuse; the isolation of nuclei in bulk and the method of fixation by freeze-drying first practised in the 1870's have both been employed once more in recent years. Such historical parallels reflect no discredit on the cytologists of the present day. That the present epoch is one of great advance in the whole subject is plain beyond any doubt. A greater awareness of historical perspective, however, would be a welcome improvement in cytological literature. It might help to dispel the illusion too often implicit in reports of fresh discoveries, that the final secrets are nearly within reach.

CHAPTER ONE

The Development of Microscopical Observation

IF ANY of us were asked for a discourse on such a topic as 'The Nature of Living Matter' we would probably begin with a series of propositions with which those who study biology are soon familiar. These basic principles may be stated briefly in this way:

(1) Living matter, in all but the smallest organisms, is nearly always divided into units, which we call cells. The great majority of cells contain a special organ, the nucleus.
(2) Cells originate only from pre-existing cells.
(3) When the cell divides, the nucleus divides first and the two nuclei which are formed are exact replicas of the original one.

On these aphorisms is based a very great deal of our modern biology; they are certainly implicit in all of its laboratory divisions. Yet the time at which all three were fully recognised is not far beyond living memory. In 1859, when the *Origin of Species* first appeared, the first of these three principles was accepted, the second had only recently been enunciated and by many was denied, and any suggestion of the third had only occurred to single and isolated observers.

Since a cell is a very small affair, it is clear that nothing of this kind of biology could begin until means were available for magnifying living objects to a sufficient extent. Yet, as we shall shortly discuss, the use of the microscope began early in the seventeenth century, and well before its end more than one observer had published drawings which showed a cellular structure. How, then, did two further centuries pass before the full significance of such observations came to be understood?

Part of the answer to this question is the technical one of the capabilities of the microscope throughout the period. For this reason we shall have to devote our first chapter to the history of the instrument. Only if we have some idea how much the early workers were able to see through their microscopes can we begin to assess their achievements and attempt to decide to what extent they were handicapped by imperfect vision, both optical and intellectual. Both of these impediments have retarded the progress of cytology in varying proportions throughout its history. At present we are certainly still conscious of the first of these limitations, though it must be left to a future historian of the subject to estimate the weight of the second factor.

Although by far the greater part of the story which these pages will attempt to tell will lie in the latter half of the nineteenth century, it is first necessary to go back to the beginnings of microscopy, if only to see the later developments in the perspective of the whole. The main facts about the history of the use of lenses as magnifiers of natural objects are soon told. Lenses have been and are used either singly or in combination. It is not possible to assign either time or place to the first employment of the single lens. The use of two lenses in conjunction to make distant objects appear nearer or small ones larger seems to have begun mainly in Holland at the beginning of the seventeenth century, though it is probable that single lenses of sufficiently short focal length to give a magnification of more than a very few diameters were not available before the middle of the century.

It was at Rome that the use of both telescope and microscope began. Duke Federigo Cesi had founded the Academy of the Lynx in 1601 (Ornstein, 1928). Its aim was the keenness of vision of that animal; Galileo became a member in 1609. Duke Federigo was interested in bees, and so this insect was chosen as the first object to be drawn at the microscope. The original plate of Francesco Stelluti (1577–1653) is at a magnification of five diameters. It was issued in 1625; one copy alone has survived. These drawings are more widely known, however, from their inclusion in a further work of Stelluti, his translation into Italian of the satires of the late Roman poet Persius Flaccus. Among them is given for good measure, but with no greater relevance, a new drawing of the weevil.

The Academy of the Lynx came to an end in the 1630's, but members of the later Accademia del Cimento such as Torricelli were also concerned with microscopes. However, sustained work with the instrument did not begin until the English Royal Society had been founded and had appointed its first curator, Robert Hooke (1635–1703).[1] Yet microscopy was only one among his many scientific interests, and his justly famous *Micrographia*, published in 1665, is by no means exclusively devoted to the 'Physiological Descriptions of Minute Bodies' of the title-page. Yet here, among so much else, is the beginning of our present subject.[2] Cellular texture is illustrated for the first time, and the word 'cell' is used in our sense of the word.

The second great figure in this field, a contemporary of Hooke and also a curator of the Royal Society, is Nehemiah Grew (1641–1712). His microscopical work is devoted wholly to plant structure, on which subject he wrote several works (Grew, 1672, 1673, 1682). In all, he published well over a hundred engravings from microscopical drawings. Many of them could well be used to illustrate a modern botanical textbook.

Both Hooke and Grew used the same type of compound microscope (*Plate I, Fig. B*), and indeed one instrument, the property of the Society, was used by each of them in turn. We

[1] Mrs Margaret 'Espinasse (1956) has earned the gratitude of all students of the history of science with her recent biography of Robert Hooke.

[2] The influence of microscopical discovery on general literature and thought in England has been traced by Nicolson (1935). The study of objects superficially insignificant or disgusting provided scope for several satires (Stimson, 1949, ch. 8). Thomas Shadwell's *Virtuoso* owed much to the *Micrographia*, and on the occasion when Hooke saw the play he was embarrassed by being recognised as the original of Sir Nicholas Gimcrack ('Espinasse, 1956, p. 150).

know a good deal about these first English compound microscopes, who made them, and how much they cost. They ranged in price from £3 upwards. The magnification was about thirty diameters. Quite a few examples have survived to this day.

In 1673, the year following the publication of Grew's preliminary *Anatomy of Vegetables Begun*, there was sent to the Royal Society a letter on several microscopical observations from a citizen of Delft in Holland, a Mr Antony van Leeuwenhoek (1632–1723), linen draper of that town. This was the first of a series of some two hundred letters concerning his researches (Dobell, p. 389) which he wrote to the Royal Society, spread over no less than fifty years. In them he went far beyond the limits set by the low-power microscopes of Hooke's pattern, and at magnifications of several hundred diameters observed details of biological structure, some of which were not seen again for over one hundred years. How, then, was it possible for this unlettered amateur to achieve such astonishing results? The answer in part is that he chose a much simpler type of microscope, one which needed infinite patience to use, but which avoided much of the optical aberrations of the contemporary compound form. Leeuwenhoek's microscopes consisted of one tiny single lens crudely mounted between brass plates (*Plate I, Fig. A*). Simple microscopes of short focal length had been used before Leeuwenhoek's time; Torricelli at the Accademia del Cimento had discovered how to make such a lens just by melting a fragment of glass into a sphere (Waller, 1684); but Leeuwenhoek himself prepared his lenses by grinding. On one occasion he took a grain of sand and ground it into a lens.

Just how Leeuwenhoek managed with his microscopes to see animalcules, bacteria, spermatozoa, and the nuclei of blood corpuscles is still not fully understood. He kept the details of his methods jealously to himself. However, it is a matter of elementary optics that the errors of simpler lenses are multiplied when they are used in series. The chief of these errors are of course spherical aberration, by which rays leaving the surface of the lens at different distances from the axis are brought to a focus at different points, and chromatic aberration, which brings light of different colours to different foci and spreads the whole spectrum round each point in the image. Only by restricting the aperture of the lenses of a compound microscope such as Hooke's to a central circle can a tolerable image be obtained. With such an instrument one does indeed see through a glass darkly. A single lens, however, can be used at a much larger aperture. It is further possible that Leeuwenhoek may have ground his lenses to an aspherical surface, which would reduce their errors still more. Again, he may well have been a person of abnormal visual acuity. Certainly the advantages of the simple microscope were recognised in his own day, especially after his letters to London began. The growth of interest in this form of the instrument at that time can be traced in Birch's *History of the Royal Society*, when references to the simple microscope then tend to increase in number. Leeuwenhoek, however, remained a unique figure; none were found to continue his work with the higher magnifications of the instrument; and Hooke as early as 1691 complained of the dearth of microscopists, which, as he said, 'are now reduced almost to a single votary which is Mr Leeuwenhoek, besides whom I hear of none that make any other use of that instrument' (Derham, 1726).

Apart from a single anonymous paper in the *Philosophical Transactions* of 1703 in which are given some figures of Protozoa and bacteria seen through a simple microscope, and which have been described by Dobell as 'amazingly good', these higher magnifications of the microscope were virtually not used after Leeuwenhoek's time, until the modern era of the subject began in the 1830's.

The miscroscope was, however, widely used during the eighteenth century, though mainly by amateurs, for whose guidance several works were produced (Baker, 1742; Ledermüller, 1764–8). Some serious students, indeed, made considerable progress in the study of small organisms, yet for all this the contemporary compound microscope was adequate. Indeed, a much better instrument could then have been produced by substituting mirrors for lenses. From Cambridge, of all improbable places at that time, had come a design for such a microscope which, had it been put into effect, would have been 'both achromatic and aplanatic up to a magnification of 300 diameters' (Nelson, 1910). The original description is to be found in the *Treatise on Optics* (1738) of Robert Smith, Master of Trinity, and Plumian Professor of Astronomy. At that time, however, the impetus to pursue further enquiry into minute bodies was not sufficiently strong for these possibilities to be explored.

As it happened, further progress was delayed until a basic improvement in the lenses of refracting telescopes had been applied to the objective of the compound microscope. It is well known in the history of optics how Newton wrongly concluded that the correction of chromatic aberration in a lens was impossible, and how this problem was ultimately solved by the combination of components of different types of glass, varying in refractive index at differing rates with the wave-length of the light transmitted. An achromatic refracting telescope was first made in 1733 by Chester Moor Hall (1703–1771), who secured for it no recognition. Some twenty-five years later the solution of the problem was rediscovered by John Dollond (1706–1761).

The same principles were applicable to the microscope, as Benjamin Martin pointed out in 1759. However, the technical problem of combining together in one unit small lenses of such steep curvature was much more difficult than had been so for the telescope. The first successful achromatic compound microscope was made about 1791 by an officer of the Amsterdam cavalry, François Beeldsnijder, but it was of relatively low magnification. This instrument has survived, and is now among a collection of historic microscopes at Utrecht, all of which are kept in working order. Measurements have been made of their performance, and a comparative account of all this data has been published (Van Cittert, 1934). Beeldsnijder's achromatic objective is well ahead of any contemporary uncorrected lens of the same power.

The earliest achromatic objectives made commercially during the first quarter of the nineteenth century in France by Selligue and in Germany by Frauenhofer consisted of separate achromatised elements which could be screwed together in various ways. The problem of spherical correction remained untouched, and only by selecting empirically the best combination of these lenses could this kind of error be reduced. Higher magnifications were obtained by adding more lenses, but their mounting aberrations soon set a limit to the practical magnification which could be achieved. In 1824 Fresnel found that only below magnifications of 200 diameters did a contemporary achromatised objective have any advantage over an uncorrected lens. A few years earlier, Jean Baptista Amici (1784–1860), who was to play a large part in later developments, turned from his attempts to improve the lenses of the microscope to further development of the reflecting instrument. A form in which an elliptical mirror was used was described by him in 1820 (*Plate I, Fig. C*).

By this time interest in some branches of microscopical study had revived, even before there had been any increase in the scope of the instrument. Plant anatomy was the first subject in which interest was renewed; Sachs tells us that

> hitherto ten or twenty years had intervened between every two works on phytotomy; but in the course of the twelve years after 1800, nearly as many publications followed one another. [Sachs, 1890, p. 256.]

Rather later, embryology began to make rapid progress under Pander and with von Baer. In both of these fields, however, there was ample scope for enquiry assisted only by very modest degrees of magnification. In Protozoology, on the other hand, Dobell (1932, p. 381) tells us that 'the first quarter of last century is almost a blank'. It may thus be said that only where improvements in the microscope were a necessity for further progress did revival wait for the appearance of the new instruments. The researches of Robert Brown (1773–1858) at this period, however, are sufficient warning against attempting to relate the renaissance of microscopy too closely to technical improvements. He may be said to have resumed where

Leeuwenhoek left off. In 1828, with a simple microscope using magnifications up to 300 diameters, he discovered the random thermal motion of small particles which is named after him and five years later he drew attention to the constant presence of an 'areola' or nucleus in the cells of flowering plants.

By this time, however, a cardinal event in the history of the compound microscope had occurred. Joseph Jackson Lister (1786–1869), father of Lord Lister, discovered that the degree of spherical aberration of an achromatic lens varies at different points along the axis, and is at a minimum at two of them, which he termed the longer and shorter aplanatic foci, and that this condition could be maintained in combining two such lenses. 'For this', as he said, 'the rays have only to be received by the front glass from its shorter aplanatic focus and transmitted in the direction of the longer correct pencil of the other glass.' This principle was laid down in a paper in the *Philosophical Transactions* of 1830. For several years previously he had been using a microscope with an objective constructed in this way, and in 1827 had published in collaboration with the physician Thomas Hodgkin (1798–1866) a paper with the title 'Notice of some Microscopic Observations of the Blood and Animal Tissues'. With this paper animal histology may be said properly to begin.

Various authors during the eighteenth and the first quarter of the nineteenth century had spoken of the substance of the organs of the animal body being made up of 'globules' and it is a matter of debate whether any of these were in fact cells. Occasionally they may have been, but usually these 'globules' were merely circles produced by optical interference in the field of a microscope of so low an aperture as to be incapable of proper resolution of the object. This verdict becomes still more probable when we study the writings of Milne Edwards, whose paper which appeared in 1823, is one of the last on globulism. Every organ of the body, according to him, was made up of globules, nearly all of which were uniformly 1/300 of a millimetre in diameter.

Hodgkin and Lister's paper in 1827 breaks cleanly with this

tradition. Each tissue which they examined had a different appearance; they gave, for instance, a description of the two types of muscle fibre. In one, 'clear and fine parallel lines of striae may be distinctly perceived, transversely marking the fibrillae, whereas in those from the middle coat of an artery no transverse striae were to be seen'. Again the shape of the human red-blood corpuscle was correctly described, but the acuity of their observation is evident even more by their failure to find within the human corpuscle the 'colourless central globule' which now we call the nucleus, which had previously been described by several authors from Leeuwenhoek onwards in the blood corpuscles of lower vertebrates. The anatomist R. D. Grainger was shown human blood through Lister's microscope, and in his *Elements of General Anatomy* (1829) makes this comment: 'My observation entirely confirmed that of Dr Hodgkin, excepting that I thought a central corpuscle could be detected' (p. 50).

Lister's work on the microscope belongs to the period of the 'Decline of Science in England' which Charles Babbage so deeply deplored in 1830. At that time there were none in this country to follow Lister's example, and so we find that in the following decades the further development of microscopy in general, and of animal histology in particular, took place mainly in continental Europe. There a great investigator, Johann Evangelista Purkinje (1787–1869) led the way, closely followed by others. Moreover, there were manufacturers of scientific instruments to provide the new improved microscopes in sufficient numbers, first for the new generation of research workers, and soon for their students in medical schools and universities. Purkinje himself, Professor of Physiology at Breslau from 1822, began work with a simple microscope; his first studies were on the anthers of flowering plants, and the next on the oocyte nucleus of the hen's egg, which is still called after him. Early in the 1830's he changed to an achromatic compound microscope and began to study ciliary movement. At that time he began to give practical classes in microscopy in his own house.

The formation of a school of workers was as important in

this field as in any other. Purkinje and his pupil Deutsch took up the study of the development of bones from the point which Clopton Havers and Leeuwenhoek had left it; and their success had a profound influence on Johannes Müller (1801–1858), Professor of Physiology at Berlin, who was no stranger to the microscope, but had hitherto considered it only of value 'in the examination of isolated particles or of thoroughly transparent textures' (Virchow, 1859). Müller too began to study the microscopical aspects of ossification. The newer version of the instrument could not then have gained a more influential advocate, nor one whose interests spread more widely into the various branches of animal biology. Among his pupils were Jacob Henle (1809–1885) and Theodore Schwann (1810–1882). The sense of adventure of which these men were conscious is expressed in a remark of Henle's which related to the time in 1834 when he and Schwann were working side by side in Müller's laboratory.

> Those were then happy days which the present generation might well envy us, when one saw the appearance of the first good microscopes from the firms of Ploessl at Vienna and from Pistor and Schieck at Berlin, which we students bought with what money we were able to save. [Frédéric, 1884, p. 13.]

Measurements made at Utrecht of the resolving power of such achromatic microscopes of the 1830's have revealed the interesting fact that their performance is just about on the same level as that of the best of Leeuwenhoek's single lenses. The new compound instruments were, of course, very much more convenient in use and were available in sufficient numbers to be widely employed (*Plate I, Fig. D*). Yet the revival of interest in the use of the microscope at that time was an even more important factor than the level of technical progress which the instrument had then reached. In each branch of microscopical biology there were figures of great importance at work during the 1830's, mainly in various European countries. If the history of any of these fields of study is traced back, one will come to some significant developments which occurred at this time. It is true that the impact of the first accurate observations upon the previous vague erroneous notions on the texture of tissues and organs gives an impression of great confusion, and this was sometimes admitted at the time. Thus William Bowman in his classical paper on muscle fibres (1840) tells us that

> The improvements which have taken place in the construction of microscopes appear to have only afforded grounds for new differences of opinion, as may be seen by the records of the last few years.

The latent possibilities of the development of the achromatic objective, however, were then very great, and were gradually achieved during the remainder of the century. Before 1830, Amici had realised that the future lay with the refracting microscope and so resumed his experiments on its development. He proceeded on lines different from those of Lister; instead of combining lenses each of which was separately corrected, he attempted to balance the errors of each component against one another. He kept closely in touch with manufacturers, particularly in Paris. By the 1840's the best of their achromatic compound microscopes had clearly overtaken the possibilities of the single lens. In this country at that time such firms as those of Ross and Powell were steadily improving their productions, but the leading British histologists seem then mainly to have used continental instruments. Goodsir at Edinburgh and Bowman at London used French microscopes, while Martin Barry chose one by Schieck of Berlin. We are told that Bowman found French microscopes considerably cheaper for class use than were those of English firms. By 1840 courses in microscopical anatomy were held in the Medical Schools at both London and Edinburgh. The ancient universities followed them at some distance. At Oxford Henry Acland began in 1845 to give lectures followed by practical illustrations, which are described in Tuckwell's lively pages in these terms:

> [The lectures] were delivered in the downstairs theatre, whence we ascended to the room above, to sit at tables furnished with

> little railroads on which ran microscopes charged with illustrations of the lecture, alternatively with trays of coffee. A few senior men came from time to time, but could not force their minds into the new groove. Dr Ogle, applying his eye to the microscope, screwed a quarter inch right through the object, and Dr Kidd, after examining some delicate morphological preparation—made answer first, that he did not believe in it, and, secondly, that if it were true he did not think God meant us to know it. [Tuckwell, 1907, p. 46.]

The resolving power of the objective of a microscope is defined as the minimum distance between two points in the field which can be separately seen. It is inversely proportional at a given wave-length to the refractive index of the medium between the object and the lenses, and also to the sine of half the angle at the apex of the largest cone of light which the front lens of the objective will usefully admit. In the further development of the achromatic objective during the nineteenth century the resolving power was improved with regard to both these quantities. As the errors of the constituent lenses were still further eliminated, it became possible for them to admit a larger cone of light without flooding the image with unwanted glare. Measurements of the resolving power of achromatic microscopes of various periods have been made, both by contemporary workers (Nobert, 1846) and in recent times (van Cittert, 1934). In the 1840's the microscopes of Purkinje, Müller, Henle and Schwann were able to resolve points just under a micron apart. An ordinary student's microscope of today with a 4-mm. objective can at best resolve points at a third of this distance. This level of performance was first reached about 1870.

The second path along which improvements in resolving power were made was by increasing the refractive index of the medium between the object and the front lens of the objective. The first 'immersion' objectives, as they were called, were used with either water or glycerin; only much later were understood the full possibilities of this method. In 1878 J. W. Stephenson pointed out that if an immersion medium had the same refractive index as that of the front lens of the objective and of the coverslip, then there would be no loss of light by reflection at these surfaces, and spherical aberration between the object and the front lens of the objective would be eliminated. Before Stephenson's paper had been published the idea had already been submitted to Ernst Abbé (1840–1908), who was at the head of the Zeiss concern, by then in the leading place among manufacturers of microscopes. In the following year there appeared the first 'homogeneous immersion' objectives. Their maximum resolution was about a quarter of a micron. A further advantage of the system was that the performance of these lenses was independent of the thickness of the coverslip, allowance for variation in which was, and still is, necessary for dry objectives of wide aperture. The introduction of this new principle is associated with a number of discoveries which depended upon this fresh advance in resolving power. Yet, however, without corresponding developments in the ancillary techniques of microscopy which fortunately came to fruition just about this time, the exploitation of the enhanced magnifications which the new immersion lenses made possible could not have followed so swiftly on their introduction.

It is not generally realised how different was the technique by which such workers as Henle, Schwann, and Bowman prepared their material from that which has since become standard. With most animal tissues the usual practice was to tease out or squash fresh material into a layer of sufficient thinness, and to study this directly under the microscope. For such objects, as everyone knows who has used an ordinary microscope on fresh material, only a very low aperture can be effectively used, and it may well be that the achromatic microscopes of the 1830's and 40's were adequate for such methods. Astonishing results could be obtained in this way in some fields—as, for example, in Bowman's studies on striated muscle fibres (1840),[1] since when, according to Barer (1948), 'remarkably little real knowledge has been added to the straightforward microscopy of muscle'. However, although the structure of

[1] Bowman's observations were aided by the use of acetic acid.

relatively uniform tissues could sometimes be adequately studied in this way, the interrelationships of different types of cell within organs was completely obscured. The botanists were more fortunate, because thanks to the nature of the cell walls in plant tissues free-hand sections could relatively easily be prepared and had been familiar from the days of Grew and Malpighi.

The class of biological reagents which we call fixatives came into use under several headings. In the first place they were used simply as preservatives for gross specimens. In 1666, Robert Boyle (1627–1691) wrote to the Royal Society 'of preserving Whelps taken out of the Dam's womb, and other *Foetus's*, or parts of them, in *spirit of wine*'. Experiments on the antiseptic properties of various substances were made by Sir John Pringle in 1750, and in 1833 J. Jacobson of Copenhagen suggested the use of chromic acid and its salts as preservatives. In microscopy, however, the main purpose for which substances of this kind were used was as 'hardening agents'. By prolonged immersion in suitable fluids animal tissues became sufficiently rigid to permit of free-hand sections being cut. The introduction of such methods was a great step forward in animal histology and embryology, and several important lines of research could hardly begin until the use of these agents was understood. Nowhere were they more necessary than in the study of the nervous system.

In 1840 Adolph Hannover described in a letter to Jacobson how a visit to Copenhagen had led him to try the effect of chromic acid on the eye and brain with great success. After this treatment it was possible to cut thin sections. The first systematic studies of the microscopy of the central nervous system in free-hand sections of hardened material began some ten years later with those of J. L. Clarke on the spinal cord of mammals. He immersed the fresh cord either in pure alcohol or in mixtures with acetic acid. Sections were cut in alcohol and were finally mounted in Canada balsam after treatment with turpentine. The plates which accompany his paper in the *Philosophical Transactions* for 1851 illustrate the possibilities which these simple methods afforded at that time to a master hand.

In embryology it was not until a few years later when Robert Remak (1815–1865) began to use hardening agents that the process of cleavage in the blastomeres of the frog's egg could be understood. The next step in the development of fixation techniques was taken with the discovery that by the use of osmic acid fine cellular detail could be preserved in a life-like form. In 1865, Max Schultze (1825–1874) published a study on the luminescent organ of the glow-worm, in which he was able to see the pattern and relationships of the tracheal end-cells in preparations treated with osmic acid.

The history of staining methods has received considerable attention in recent years; Conn (1928, 1930) and Baker (1943) have described the earliest instances of the use of dyes on microscopical preparations, and later developments have been traced in a series of papers by Conn and others. Carmine was used independently by several workers in the 1850's (Corti, 1851; Osborne, 1857; Gerlach, 1858) as in the following decade were haematoxylin and the first anilin dyes (Beneke, 1862; Schweigger-Seidel, 1865; Conn, 1930). Usually at first these substances were used somewhat crudely to give a general coloration, and methods for differential staining of the various components of cells and tissues were not developed until the 1870's.

During this decade a number of workers, of which Walter Flemming (1843–1915) was the most important, refined the techniques of fixation and staining to a sufficient degree to enable them to study the details within the cell and its nucleus. Modern cytology begins at this time; its first major achievement was the understanding of the changes which lead to the division of the nucleus. Thus by the time that the oil-immersion lens had made its appearance there were microscopical preparations which could take advantage of its possibilities.

Corresponding refinements had by then been made in the methods of preparing sections of biological material. Again we find that the first halting steps in this direction were made before 1860, and that adequate methods were not developed for more than another decade. Various substances were suggested as suitable supports for a specimen during the cutting of sections, and were widely used during the 1870's. Of such, Foster and Balfour's *Elements of Embryology* (1874) lists paraffin wax, mixtures of white wax and oil, spermaceti and gum. Infiltration of the specimen by the molten substance before embedding it in the cooled solid was not introduced until a group of workers at the Zoological Station at Naples, faced with the task of studying the structure of the delicate larval forms of marine invertebrates, greatly refined the technique of embedding and the preparation and manipulation of sections (Mayer, 1880). Visiting English embryologists, chief among which was F. M. Balfour (1851–1882), introduced such methods into this country. The technique in which each section was handled separately was even then still extremely laborious; Shipley has recorded how in 1880 the afternoons of half a term were needed to section a small Amphioxus (Shipley, 1924, p. 162). However, within the next two years two members of the same school, Caldwell and Threlfall (Threlfall, 1930), discovered that under appropriate conditions consecutive sections could be welded together to form a continuous ribbon and were able to use this as the basis for an automatic machine for cutting sections. The first model was driven by a small water motor, and is still to be seen in the Cambridge Laboratory of Zoology. By the time that it was first at work Balfour was no longer there to take advantage of it. Embryology in Cambridge has never recovered from his untimely death.

Thus by the early 1880's cytologists were in command of a number of technical facilities which had been separately developed in the preceding years. Refined methods of staining, the new techniques of accurate section cutting, and the oil-immersion lens were all first available much at the same time.

One further refinement of the light microscope was, however, yet to come. Early in the 1870's, Ernst Abbé (1840–1908) was aware that further correction of the microscope objective would be possible if the choice of material for the lenses was not restricted to glasses of the crown and flint series alone (Hughes, 1957). In these, the dispersion, or rate of change in refractivity with wave-length, is proportional to the refractive index itself, whereas other relationships between these quantities, he realised, would greatly enlarge the scope of the designer of objectives. Early in the 1880's Abbé joined with Otto Schott, a glass manufacturer, in experiments on adding various chemical elements such as boron and phosphorus to the silicate basis of glass. By 1886 they had produced their Jena glasses with entirely novel characteristics.

The improved lenses which these new materials made possible were called 'apochromatic', for they eliminated the residual chromatic aberration, the secondary spectrum, of the achromat (Abbé, 1886). They were first available in the same year as were the new Jena glasses. With the apochromatic immersion objective of N.A. 1·4 the resolution of the microscope for white light reached a limit at which it has remained for nearly seventy years. Early in this period cytology advanced from the study of the whole nucleus to that of the individual constituent chromosomes. German cytologists were using apochromatic objectives within a few years of their introduction (Hermann, 1891; Hertwig, 1890; Flemming, 1891), although elsewhere they were taken up rather more slowly. Over the past seventy years the main stimulus to the progress of chromosome cytology has come not so much from enhanced technical possibilities within the subject itself but from its convergence with the science of heredity. The history of this aspect of the subject we shall study in later pages.

Within recent years, however, there have been several exciting new developments in microscopy. From complete dependence on preparations which have been fixed, sectioned, and stained, we are now able to go back to the study of fresh material, but with microscopes embodying new principles in optics which can reveal a living cell with all the contrast and resolution which have hitherto been attainable only with stained preparations.

These new microscopes work on the differences in refractive index within the object and, in amplifying these, give contrast in the image by the introduction of destructive interference. Dark areas in the picture are where rays opposite in sign have cancelled each other out, and may be produced either by points in the object of either higher or lower refractive index than their surroundings. Two forms of such microscopes are in use; one depends on a relatively simple principle invented by the Dutch optician F. Zernicke some twenty years ago. Interest in its application to the microscope was not aroused, however, till at the end of the late war it had acquired the status of a German technical secret, so accustomed had microscopists become to the conventional methods of fixation and staining. A new form of interference microscope of still greater sensitivity has been introduced within the last two years.

Within the last decade there have also been developments of great interest in ultra-violet and in polarising microscopes. Even more important, however, has been the transference of the limits of resolution to far lower orders of magnitude by the use of electron beams in place of light rays. The electron microscope, which has been known for over twenty years, has only recently begun to provide information about the constitution of the cell. Again, as was true a century ago, the available optical resources could not be adequately exploited until the ancillary microscopical techniques were available. The electron microscope did not become a cytological instrument until the ultra-microtome was developed, capable of cutting sections of extreme thinness. The discovery that broken glass can be sharp enough to cut sections nearly one thousandth the thickness of those commonly used with the light microscope is largely responsible for bringing the technology of electron optics into cytology. As in the previous century, animal histology again proved to be one of the more difficult fields to be studied by the new methods.

We see, therefore, that progress in magnifying power and the possibility of resolution of fine details are by no means the only factors which have promoted further researches into the structure of minute bodies.

A field, however, in which subsidiary techniques have played only a minor part is the study of the spermatozoa. It will be appropriate, then, to illustrate the progress in microscopical enquiry which we have so far been discussing by reference to figures of these objects published during the last two and a half centuries, more particularly as Leeuwenhoek himself was celebrated in his own day more for his descriptions of the 'spermatic animalcules' than for any other of his discoveries. Leeuwenhoek's drawings of spermatozoa were first published in the *Philosophical Transactions* in 1679, to illustrate his letter of March 18, 1678, to Nehemiah Grew, who was then the secretary of the Society (*Plate II, Fig. A*). Of these figures, the first four are stated to be of human sperms in the recent van Rijnbeck edition of Leeuwenhoek's works, though they have in the past been ascribed to the rabbit (Cole, 1930, p. 13; Meyer, 1939, Fig. 35). In the text of the letter they are described as from 'male semen'.

The last four figures are of sperms of the dog. In both sets only the first, namely Figures 1 and 5, is from living material. Certainly Figure 1 is much more like a human sperm than are the others, and it is on this that Leeuwenhoek's achievement in this field is to be assessed. It is still mysterious how a drawing no less inexact than this, at a magnification of over 2,000 diameters, could have been made at that time. Certainly no better appeared for another century and a half, as we may see by comparison with the figures of the later authors. One of these, M. F. Ledermüller, also used a simple microscope. His drawings were published in 1758; they represent sperms of man, the perch, the frog, and of a snake and a tortoise. The original magnifications may be estimated at about × 400 from the size of the head of the human sperm. Apart from the usual defect of too short a tail, the first four have some resemblance to their subjects, but his Figure 61 is quite unlike the sperm of a tortoise or indeed of any other animal (*Plate II, Fig. B*). In 1821, Prévost and Dumas studied spermatozoa with an uncorrected compound microscope. Their figures represent no advance on those of the eighteenth century; they were reproduced at the fantastic magnification of 3,000 diameters, at which the maximum resolution of their microscope is probably represented by points some 15 millimetres apart (van Cittert, 1934; Hughes, 1955). With this figure in mind we can study these figures of Prévost and Dumas. They are of the sperms of various mammals to which none of them bear any particular resemblance (*Plate III*). For example, there is no hint of the very characteristic hook-shaped head of that of the mouse.

The achromatic microscope was first used in this field in 1837, when two papers appeared, one by Felix Dujardin (1801–1862) and the other by Rudolph Wagner. Their figures of spermatozoa were both reproduced at about 800 diameters (*Plate IV, Figs. A and B*). In comparing the work of these authors we see that Wagner restricted himself to outline drawings, while there is some detail to be seen within Dujardin's figures. For instance, in the human sperm the latter saw both a middle-piece and some cytoplasmic remnants. He misinterprets, however, the shape of the head of the mouse sperm, which is accurately drawn by Wagner. Both authors also described amphibian sperms, Wagner in a later section of the same paper and Dujardin in a further publication in the following year. Here the French author clearly wins, for he recognised the existence of the undulating membrane attached to the sperm tail in *Triton,* while Wagner drew only the merest traces of this structure.

Of the representations of spermatozoa by Dujardin and Wagner, it may be said that these authors drew only what they were able to see. Others, however, were so strongly influenced by the conviction that the spermatozoa were foreign and parasitic organisms that they were able to identify a visceral apparatus within them. Such at this time were Valentin (1839), Gerber (1842), and Pouchet (1847) (*Plate IV, Figs. C, D, and E*). All three were familiar with the best microscopes of the day, but Pouchet provides the chief example of the failure of optical progress by itself to displace antecedent conviction; more especially since six years previously both Kölliker (1841) and Dujardin (1841) had described the actual mode of origin of the spermatozoa.

Pouchet's figures are found in the atlas accompanying a work entitled *Théorie Positive de l'Ovulation Spontanée* (1847). He is better known as an opponent of Pasteur in the later controversies over spontaneous generation. He had, as Cole (1930, p. 33) tells us, an 'ardent temperament, to which he surrendered without a struggle'.

Two years before Pouchet's atlas had appeared a rather different work had been published in Paris which also included representations of spermatozoa. The *Cours de Microscopie* . . . of Alphonse Donné is free of all personal error, for its illustrations were based on photomicrographs. Donné and his collaborator, the young Léon Foucault, were among the earliest to use the photographic process of Daguerre (Moholy, 1939) and by far the first ever to combine camera and microscope. Their apparatus is described in the atlas. Illumination was by sunlight whenever possible, but otherwise they used 'L'incandescence du charbon sous l'influence d'un courant éléctrique'. The exposure on their iodised silver plates was no more than 4–20 seconds at magnifications of 200–400 diameters. The results at that time were necessarily reproduced as lithographs, but their accuracy is clearly evident, for we can easily recognise distinctive features in the spermatozoa of several mammals. The hook-shaped head in the mouse and the thick middle-piece in the bat are both evident (*Plate V*).

It was not before another twenty years that we come to a further contribution to the microscopy of the spermatozoa which we can recognise as a distinct advance on the work of Donné. In 1865 a paper was published by F. Schweigger-Seidel in this field. He gave a plate of drawings of the spermatozoa of a number of vertebrates, studied by means of a water-immersion objective. Several methods of staining were used in the course of this work, which is one of the very first examples of a histological study in which aniline dyes were employed. Schweigger-Seidel was the first to recognise the existence of a distinct zone, the middle-piece, between the head and the tail of the spermatozoon (*Plate VI, Fig. A*).

As we shall find in other branches of microscopical enquiry, from the later 1870's there is then a steady increase both in the number and the quality of the constituent contributions. The effect of the homogeneous immersion objective is seen particularly well in studies on the mouse spermatozoon, which is one of the largest among the Mammalia. The spiral sheath of the middle-piece, which we now know consists of mitochondrial filaments, came into view when first examined with the aid of these lenses. Thus, for example, Olaf Jensen in 1879 saw no more with a water-immersion lens than Schweigger-Seidel had done fourteen years before, but in his later paper of 1887 Jensen was then able to draw the spiral filament with considerable accuracy (*Plate VI, Fig. D(i)*). This structure was seen a few years earlier by some of the first users of the oil-immersion lens (Leydig, 1883; Brunn, 1884), but was first described as a 'cross-striation' (*Plate VI, Figs. B and C*).

The first to use an apochromatic objective on spermatozoa was the English microscopist E. M. Nelson. In two papers, in 1889 and 1892, he described his observations on the human sperm, the illustrations of which up to that time had been, as he said, 'only up to the microscopy of early achromatic days'. In these papers Nelson corrected earlier errors but introduced some new ones. It was, however, reserved for Gustav Retzius (1842–1919) to do full justice to the possibilities of the apochromatic objective in this field. Retzius was a Swedish anatomist and the son of a distinguished microscopist. His interests were very wide; he was expert in both anthropology and histology, and moreover wrote biographies of a number of scientists. His researches in cytology were published under the heading of 'Biologische Untersuchungen', first as volumes with articles by himself and his colleagues, but later as single folios on his own observations. These were distributed as gifts among biologists in many countries—a practice which his ample means made possible. Of Retzius' microscopical investigations, none are more distinguished than his later studies on the spermatozoa

(Retzius, 1909). With them we reach the apex of the possibilities of the light microscope; for no finer tribute to the apochromatic objective has ever been paid (*Plate VI, Fig. E*). The increased scope of the microscope made possible by this lens can at once be appreciated by comparison of this work with Retzius' earlier study on spermatozoa in 1881, one of the first in which an achromatic oil-immersion lens was used.

The investigation of the structure of the spermatozoa did not advance beyond the level which Retzius had gained for more than thirty years afterwards. Thus even within an era of such rapid development in general biology as this century there was a halt in the progress of this particular field longer than any since the eighteenth century. This interval came to an end in the early 1940's with the entry of the electron microscope into biological research; yet even then progress was only slowly resumed. All but the smallest biological objects while intact are too opaque for the electron beam to penetrate them, and so some technique of partial disintegration was found necessary. With spermatozoa, treatment of the fresh material with distilled water was found to have the effect of splitting the tail into its constituent elements; in the sperm of *Arbacia* Harvey and Anderson (1943) found that the tips then 'resemble frayed ends of rope unwrapped into separate strands'. Yet, remarkably enough, the individual fibrils which are revealed by treatment of this kind had already been seen by Jensen in 1887 with the light microscope over half a century before (*Plate VI, Fig. D(ii)*).

Numerous details of the structure of spermatozoa were, however, gradually revealed by the earlier studies with the electron microscope. It was found that the fibrils of the tail, which are the actual contractile elements responsible for its activity, are precisely eleven in number over a wide range of species. Except at the extreme tip of the tail, they are surrounded by a spiral sheath.

It was not, however, until the advent of the ultra-microtome that the relations of these structures in the intact sperm could be adequately studied. The technique of preparing extremely thin sections of minute biological objects has now advanced so far that within the last few years Dr Bradfield of Cambridge has succeeded in photographing at a magnification of some tens of thousands of diameters an actual transverse section through a sperm tail (Bradfield, 1954).

With this achievement this brief review of the study of the spermatozoon and of the capabilities of microscopes can fittingly be brought to a close. It may be said that in few other branches of cytology has progress so closely followed on the heels of the ever-widening technical capabilities of the instrument. When, however, as in the succeeding pages we consider and attempt to assess the development of other branches of research throughout their history, we must always bear in mind what could actually be seen through the contemporary microscopes at each period. The history of cytology, or of anything else, can by no means be deduced from the study of one single constituent factor; yet it is with the visual information presented to the eye of each investigator that our enquiries must begin.

LITERATURE FOR CHAPTER ONE

ABBÊ, E. (1886). Sitzber. des Jen. Gesell. f. Med. und Naturwiss., p. 107.

ACKERKNECHT, E. H. (1953). Rudolf Virchow, Doctor, Statesman, Anthropologist. Univ. Wisconsin Press.

AMICI, J. B. (1820). Edinburgh Phil. J. 2, 135.

ANON, (1703). Philos. Trans. Roy. Soc. London. 23, p. 1357.

BABBAGE, C. (1830). Reflections on the Decline of Science in England. London.

BAKER, H. (1742). The Microscope Made Easy. 1st ed. London.

BAKER, J. (1943). Quekett Micro. Club. Vol. 1, Ser. 4, p. 256.

BARER, R. (1948). Biol. Rev. Camb. Phil. Soc. 23, 159.

BENEKE, F. W. (1862). Correspbl. d. Ver. f. gemeinsch. Arbeiten. 59, 980.

BERRES, J. (1837). Anatomie der microskopischen Gebilde des menschlichen Körpers. Wien.

BIRCH, T. (1746). History of the Royal Society of London. 4 vols. London.

BOWMAN, W. (1840). Philos. Trans. Roy. Soc. London. Vol. 130, 457.

BOYLE, R. (1666). Philos. Trans., Roy. Soc., London. Vol. I No. 12, p. 199.

BRADFIELD, J. R. G. (1954). Quarterly J. Micro. Science. 94, 351.

BROWN, R. (1828). Edinburgh New Phil. Journ. 5, 358.

BRUNN, A. VON. (1884). Archiv. f. mikro. Anatomie. 23, 108.

CITTERT, P. H. VAN (1934). Descriptive Catalogue of the Collection of Microscopes in the Utrecht University Museum. Groningen.

CLARKE, J. L. (1851). Philos. Trans., Roy. Soc., London, p. 607.

COLE, F. J. (1930). Early Theories of Sexual Generation. Oxford.

CONN, H. J. (1928). The History of Staining. The Pioneers . . . Stain Technology. 3, 1.

CONN, H. J. (1930). The History of Staining. Anilin Dyes in Histology. Ibid. 5, 3.

CORTI, A. (1851). Zeit. f. wiss. Zoologie. 3, 109.

DARWIN, C. (1859). The Origin of Species. London.

DERHAM, W. (1726). Philosophical Experiments and Observations of the late Eminent Dr Robert Hooke. . . . London, p. 261.

DISNEY, A. N., HILL, C. F., and WATSON BAKER, W. E. (1928). Origin and Development of the Microscope. London.

DOBELL, C. (1932). Antony van Leeuwenhoek and his Little Animals. London.

DOLLOND, J. (1758). Philos. Trans., Roy· Soc., London. 50, 733.

DONNÉ, A. (1845). Cours de Microscopie. Paris.

DUJARDIN, F. (1837). Annales Sci. nat. (Zool.). Ser. 2, 8, 291.

DUJARDIN, F. (1841). Historie Naturelle des Zoophytes. Paris.

'ESPINASSE, M. (1956). Robert Hooke. Heinemann, London.

FLEMMING, W. (1891). Archiv. f. mikro. Anat. 29, 389.

FOSTER, M., AND BALFOUR, F. M. (1874). Elements of Embryology. Pt. I. London.

FRÉDÉRICQ, L. (1884). Théodore Schwann, Sa Vie et ses Travaux. Liège.

FRESNEL, A. J. (1824). Ann. Sci. nat. 3, 345.

GERBER, F. (1842). Elements of Anatomy with Notes by G. Gulliver (trans.). London.

GERLACH, J. (1858). Mikroskopische Studien aus dem Gebiete der menschlichen Morphologie. Erlangen.

GRAINGER, R. D. (1829). Elements of General Anatomy. London.

GREW, N. (1672). The Anatomy of Vegetables. London.

GREW, N. (1673). An Idea of a Phytological History. Propounded. . . . London.

GREW, N. (1682). The Anatomy of Plants. London.

HANNOVER, A. (1840). Arch. f. Anat. Physiol. wiss. Med., p. 549.

HARVEY, E. B., AND ANDERSON, T. F. (1943). Biol. Bull. Wood's Hole. 85, 151.

HERMANN, F. (1891). Archiv. f. mikro. Anat. 37, 569.

HERTWIG, O. (1890). Arch. mikro. Anat. 36, 1.

HODGKIN, T., AND LISTER, J. J. (1827). Phil. Mag. 2, 130.

HOOKE, R. (1665). Micrographia. London.

HOOKE, R. (1691). Dr Hooke's Discourse concerning Telescopes and Microscopes in Derham, W. (1726), p. 257.

HUGHES, A. F. (1955). J. Roy. Micro. Soc. 75, 1.

HUGHES, A. F. (1957). J. Roy. Micro. Soc. 76, 47.

JACOBSON, J. (1833). Edinburgh New Phil. Journ. 15, 157.

JENSEN, O. S. (1879). Die Structur der Samenfäden. Bergen.

JENSEN, O. S. (1887). Archiv. f. mikro. Anat. 30, 379.

KÖLLIKER, A. (1841). Beiträge zur Kenntniss der Geschlechtsverhaltniss. Berlin.

LEDERMÜLLER, M. F. (1758). Versuch zu einer grändlichen Vertheidigang derer Saamenthiergen. . . . Nürnberg.

LEDERMÜLLER, M. F. (1764–8). Amusement microscopique. 3 vols. Nürnberg.

LEEUWENHOEK, A. van, (1679). Philos. Trans. Roy. Soc. London. 12, 1040

LEYDIG, F. (1883). Untersuchungen zur Anatomie und Histologie der Thiere. Bonn.

LISTER, J. J. (1830). Philos. Trans. Roy. Soc. London. 130, 187.

MARTIN, B. (1759). New Elements of Optics. London.

MAYER, P. (1880). Mitt. d. Zool. Stat. Neapol. 2, 1.

MEYER, A. W. (1939). The Rise of Embryology. California and Oxford.

MILNE EDWARDS, H. (1823). Mémoire sur la structure élémentaire des principaux tissus . . . des animaux. Paris.

MOHOLY, L. (1939). A Hundred Years of Photography. Penguin, London.

NELSON, E. M. (1889). J. Quekett Micro. Club. N.S. 3, 310.

NELSON, E. M. (1892). J. Quekett Micro. Club. N.S. 4, 264.

NELSON, E. M. (1910). J. Roy. Micro. Soc., p. 427.

NICOLSON, M. (1935). Smith Coll. Studies in Mod. Languages. 16, No. 4. 92 pp.

NOBERT, F. A. (1846). Ann. Phys. Leipzig. 67, 173.

ORNSTEIN, M. (1928). The Role of Scientific Societies in the Seventeenth Century. Chicago.

OSBORNE, S. G. (1857). Trans. Micro. Soc. London. 5, 104.

POUCHET, F. A. (1847). Théorie Positive de l'ovulation Spontanée. Paris.

PRÉVOST, J. L., AND DUMAS, J. B. (1821). Mém. Soc. Phys. Genève. 1, 180.

PRINGLE, J. (1750). Philos. Trans. Roy. Soc. London. No. 495, p. 480.

PURKINJE, J. E. (1825). De cellulis antherarum fibrosis . . . Bratislava.

PURKINJE, J. E. (1830). Symbolae ad ovi avium. . . . Leipzig.

PURKINJE, J. E. (1834). Arch. Anat. Physiol. wiss. Med., p. 385.

PURKINJE, J. E. (1835). De phaenomeno . . . motus vibratori continui in membranis . . . Bratislava.

REMAK, R. (1858). Archiv. f. anat. u. Physiol. wiss. Med., p. 178.

RETZIUS, G. (1881). Biol. Untersuchungen. 1. Stockholm.

RETZIUS, G. (1909). Biol. Untersuchungen. N.F. Bd. 14, 11–21. Stockholm.

RIJNBECK, G. VAN (1929). Collected letters of A. van Leeuwenhoek, edited, illustrated, annotated by a Committee of Dutch Scientists, Amsterdam.

SACHS, J. (1890). History of Botany 1530–1860. Trans. H. E. Garnsey, Oxford.

SCHULTZE, M. (1865). Arch. f. mikro. Anat. 1, 124.

SCHWEIGGER-SEIDEL, F. (1865). Arch. f. mikro. Anat. 1, 309.

SHIPLEY, A. (1924). Cambridge Cameos. London.

SMITH, R. (1738). A Complete System of Opticks. 2 vols. Cambridge.

STELLUTI, F. (1630). Persio tradotto inverso scioto e dichiarato da F. Stelluti. Rome.

STIMSON, D. (1949). Scientists and Amateurs. London.

STEPHENSON, J. W. (1878). J. Roy. Micro. Soc. 1, 51.

THELFALL, R. (1930). Biol. Rev. Camb. Phil. Soc. 5, 357.

TUCKWELL, W. (1907). Reminiscences of Oxford. London.

VALENTIN, G. G. (1839). Nova Acta phys.-med. Acad. Leop. 19, 237.

VIRCHOW, R. (1859). Johann Müller, An Eloge. . . . Trans. A. M. Adams, Edinburgh.

WAGNER, R. (1837). Abh. Akad. Wiss. 2, 381.

WALLER, R. (1684). Essayes of Natural Experiments made in the Academie del Cimento . . . Englished by R. W. London.

ZERNICKE, F. (1935). Z. tech. Phys. 16, 454.

CHAPTER TWO

Recognition of the Cell, and the First Theories of its Formation

WE MUST now go back to the point where we began, and describe in some detail the earliest microscopical observations on cells. We start with Robert Hooke, but the story begins a year before the *Micrographia* was published, for in John Evelyn's *Sylva* (1664, p. 96) are some observations on petrified wood, contributed by Hooke. Among its points of resemblance to modern timber was that

> all the smaller and (if I may call those which are only to be seen by a good glass) microscopical pores of it appear (both when the substance is cut and polished transversely and parallel to the pores) perfectly like the microscopical pores of several kinds of wood. . . .

Observation XVII in the *Micrographia* continues these findings and is illustrated with sections. The following section has the heading 'Of the Schematism or Texture of Cork', but extends some way beyond the consideration of this substance alone. The shape and size of the 'pores or cells' of cork are described, and figures illustrate their different appearances in transverse and longitudinal section (*Plate VII, Fig. A*). Hooke continues:

> Nor is this kind of texture peculiar to cork only, for upon examination with my Microscope, I have found that the pith of an Elder, or almost any other tree, the inner pulp or pith of the Cany hollow stalks of several other vegetables: as of Fennel, Carrets,

Daucus, Bur-docks, Teasels, Fearn, some kinds of Reeds etc. have much such a kind of *Schematisme*, as I have lately shewn that of Cork.

Hooke observed that these cells usually contained fluid, and discusses how far they may be in communication:

> But though I could not with my Microscope, nor with my breath, nor any other way I have try'd, discover a passage out of one of those cavities into another, yet I cannot thence conclude, that therefore there are none such, by which the *succus nutritus*, or appropriate juices of vegetables, may pass through them; for, in several of those vegetables, whil'st green, I have with my Microscope, plainly enough discover'd these cells or Pores fill'd with juices, and by degrees sweating them out.

He then throws out the suggestion that there may be a circulation of fluids in plants as well as in animals, and that

> it seems very probable that Nature has in these passages, as well as in those of animal bodies, very many appropriated instruments and contrivances, whereby to bring her designs and end to pass, which 'tis not improbable, but that some diligent observer, if helped with better *Microscopes*, may in time detect.

Finally he suggests that the behaviour of 'sensitive Plants, wherein Nature seems to perform several animal functions with the same *Schematism* or *Organization* that is common to all vegetables' may be an indication of similar modes of action in both plants and animals.

In the *Anatomy of Plants* (1682) (*Plate VII, Fig. B*) Nehemiah Grew uses the term 'pores' for the vessels within the wood, and speaks of the cells of the pith and parenchymatous regions as 'bladders'. The appearance of the whole parenchyma he first compares with the 'froth of beer' (p. 64) but later exchanges this analogy for another. On page 76 he tells us

> That the *sides* by which the aforesaid *Bladders* of the *Pith* are circumscribed, are not meer *Paper-skins* or rude *Membranes*, but so many several Ranks or Piles of exceedingly small *Fibrous Threds*; lying, for the most part, evenly one over another, from

> the bottom to the top of every *Bladder*; and running cross, as the *Threds* in the weavers *Warp*, from one *Bladder* to another, which is to say, that the *Pith* is nothing else but a *Rete mirabile*, or an Infinite Number of *Fibres* exquisitely small, and admirably complicated together. [*Plate VII, Fig. C.*]

Hooke and Grew thus approach the problem of cellular structure in plants from different angles. Hooke poses the question of the functional interrelations of cells, while Grew is concerned only with the structural pattern of the whole tissue; had Hooke gone any further he would have been the first cell physiologist.

There, together with the studies in plant histology of Leeuwenhoek and Marcello Malpighi,[1] the subject remained until early in the nineteenth century. During the revival of interest in plant structure, of which mention was made in previous pages, several botanists formed the idea that cells were not just the spaces between a network of fibres but that they were separate and separable units. G. R. Treviranus in 1805 described how he had been able to isolate them from one another by teasing a section through a bud of *Ranunculus*; two years later D. H. F. Link suggested that there could be spaces between cells where, as in the pith of *Datura*, the cell walls appeared as double lines. Link also made the significant observation that in some instances a cell with coloured sap is surrounded by other cells with untinted contents. In 1824 H. J. Dutrochet (1776–1847) described how the cells of the pith of *Mimosa* can be separated from one another by boiling in nitric acid.

The main impediment to further understanding of the nature of the plant cell at this time, and for some decades to come, was that imposed by the use of the word 'cell'. The essential feature of cells in either honeycombs or prisons is their walls; and while the plant cell was reckoned to be only a wall enclosing a fluid, no comparison was possible with any structural units of the animal body.

The first observations on the microscopic structure of animals were on the corpuscles of the body fluids. Jan Swammerdam

[1] The achievements of Grew and Malpighi in this field have been compared by Arber (1942).

(1637–1680) probably saw them first, but his *Biblia Naturae* (1737–8) was not published until nearly sixty years after his death. Marcello Malpighi (1628–1694) was the first to make any reference to corpuscles, but confused them with fat globules (Baker, 1948). Leeuwenhoek from time to time described observations on blood. In 1674 he remarked that it consists of small round globules driven through a crystalline humidity or water. Twenty-six years later he described the circulation through the capillaries of the fins of 'butts', the fry of flat-fishes such as plaice and flounders, and in the same letter gives an account of the oval particles of the blood of a salmon, as seen when a drop was spread 'on a very clean glass'. Here we find the first reference to a nucleus in Leeuwenhoek's remark that in those oval particles which lay flat there was 'a little clear sort of light in the middle, larger in some than in others'. This is shown very clearly in the accompanying figures (Leeuwenhoek, 1702) (*Plate VIII, Fig. A*).

The next major contribution to the study of blood corpuscles is in the third part of William Hewson's *Experimental Enquiries*, a 'Description of the Red Particles of the Blood', published posthumously in 1777. Hitherto the human red corpuscles had been regarded as spherical, from the time when Leeuwenhoek had referred both to them and to the fat droplets of milk as 'globules'. Hewson saw that in blood diluted with serum the corpuscles were flat, but that in water they became spherical; they again resumed the original shape on the addition of a drop of a neutral salt. Hewson further examined the red corpuscles of a number of vertebrates and illustrated their comparative sizes as they appeared through a lens of $\frac{1}{23}$ in. focus. Hewson's use of the simple microscope, however, did not prevent him from falling into the error of putting nuclei into all his mammalian corpuscles.

The first denial that the human red corpuscle was nucleated was due, as we saw in the previous chapter, to Hodgkin and Lister. This conflict of opinion gave rise to some scepticism about the value of microscopical observations in general (Bostock, 1836). Had Hodgkin and Lister looked at blood corpuscles both from man and from a lower vertebrate, a good deal of confusion would have been avoided. It was not till 1841 that E. H. Weber made this comparison. He pointed out how readily the central area of the mammalian corpuscle could be mistaken for a nucleus. At different focal levels it could appear either lighter or darker than its surroundings.

The first observation of the nucleus in an adult animal cell, other than in a blood corpuscle, was made in 1781 by Felix Fontana, who, in a book mainly on the viper and its venom, includes at the end a section on miscellaneous observations with a microscope, among which is one on the slime from the skin of an eel. Within this substance he saw globules which were epithelial cells (*Plate VIII, Fig. B*), and inside them again he saw an oviform body, the nucleus. A spot (une tache) within this may well have been the nucleolus.

It was not, however, until the 1830's that the microscopical structure of living organisms begins to attract the attention of more than the single and isolated observer. Within that decade both botanists and zoologists came to recognise the existence of a basic correspondence in minute structure in both plants and animals. This era is generally regarded as that of the foundation of the cell-theory, the two names which are usually most prominently associated therewith being Theodore Schwann (1810–1882) and M. J. Schleiden (1804–1881). In recent years there has been considerable discussion about the ideas on the texture of animal tissues which were current at this period, and by now certain misconceptions seem to have been cleared away. In tracing the origins of cellular theory, we find that not only is the word 'cell' at first given a meaning which we no longer recognise, but also when structures which we now regard as cells were discovered they were often called by various other names which now have no special significance.

In the first place the expression 'cellular tissue', which is used by several authors at the beginning of the nineteenth century from Lamarck (1809) onwards (Gerould, 1922), is equivalent to what we now term 'areolar connective tissue'

(Wilson, 1944; Baker, 1948), the spaces within which were termed 'cells' by writers of that time. This usage persisted even as late as 1840, for it is to be found in the third edition of P. M. Roget's Bridgewater Treatise, where, under the heading of 'Animal organisation', cellular texture of this kind is described, in contrast to a wholly different account of 'Vegetable organisation' which is in terms of cells as we now understand them.

The use of the word 'globule' for microscopical elements in animal tissues from Malpighi onwards (1665; Baker, 1948) has given rise to even greater confusion. In the preceding chapter (p. 8, above) there was mentioned the probable source of many of these 'globules' in the diffraction circles within the image of an imperfect microscope. It is not always certain, however, how near some 'globulists' came to seeing actual cells. In particular Rich (1926) has argued that Dutrochet (1824) recognised the cellular nature of a variety of tissues. It is true that wholly modern views can be read into quotations from this work. However, Dutrochet's work belongs still to the era of the globulists, as Wilson (1947) has shown by comparing a plate of microscopical drawings from the French author with one from Theodore Schwann. A great divide separates these two investigators. Dutrochet used a simple microscope, though of low magnification. He is just able to make out the cross-striations of an insect muscle fibre, but his drawings of brain tissues of *Helix* and of the frog carry no conviction.

When the decisive period of the 1830's is reached, a new era of microscopical study then begins. The discoveries of the period in this field may be said to fall into these three divisions:

(1) Recognition of the nuclei within the cells of plants, and in the earliest stages of animal development.
(2) The discovery of cells within the tissues of animals.
(3) Recognition of the nature of the primary living substance and its first designation by the word 'protoplasm'.

Although these lines of enquiry were made to converge during this decade, yet it was a false theory of the formation of new cells by which they were drawn together, the influence of which was not thrown off for more than a score of years afterwards. It is thus necessary in studying the history of cytology in these years to view each single item of discovery against the general background of the biological thought of the time and to realise how slowly the full significance of so many individual observations became clear. The growth of this knowledge was influenced largely by those who could generalise from the observations of others, and, as we shall see, wrong conclusions, once accepted, could retard the development of the subject for many years.

Robert Brown (1773–1858), to whom the recognition of the nucleus of the plant cell is mainly due, had both acute powers of observation and the ability to draw correct conclusions from the work of others. In his paper of 1833 on the reproductive organs of Orchideae and Asclepideae, in which he recognised the presence of a nucleus in the cells of a number of flowering plants, we find these words:

> The few indications of the presence of this nucleus, or areola, that I have hitherto met with in the publications of botanists are chiefly in some figures of epidermis in the recent works of Meyer and Purkinje, and in one case in M. Adolphe Brogniart's memoir on the structure of leaves. But so little importance seems to be attached to it that the appearance is not always referred to in the explanation of the figures in which it is represented.

Brown's paper may be said to have established the concept of the nucleated cell as the unit of structure in plants. Only later was this stage reached in the study of the tissues of animals for several reasons, mainly because of the great diversity of their cells. In consequence of this variety of shape, separate names became attached to the various forms of cell and nucleus. In 1830, for instance, Purkinje described the relatively enormous nucleus of the ovarian egg of the hen and called it the 'germinal vesicle' (*Plate IX, Fig. E*). Within a few years other workers, following von Baer's discovery of the mammalian ovum, described germinal vesicles in the eggs of a number of mammals (Coste, 1833; Bernhardt, 1834). The idea of the egg as a cell and the germinal vesicle as a nucleus, however, developed only slowly in the following decades.

Within the tissues of the adult animal, recognition of cells was naturally easier where their shape was similar to that of plant cells. Such, for instance, was the epithelium of the choroid plexus, in which the arrangement of cells and nuclei suggested to Valentin in 1836 a close comparison with the epidermis of a plant. A still more powerful analogy, however, was found in stratified squamous epithelium. Henle in 1837 found that not only were the individual cells similar in appearance to those of the parenchyma of plants, but also their growth proceeded along the same lines. The smallest and youngest cells were at the base of the epithelium in the germinative layer, while older cells progressively move towards the surface and expand in volume.

This expansion of growing and developing cells in an animal tissue seemed of special significance to these authors because at that time skin was regarded merely as an exudate from the blood which subsequently hardens. In those days the growth of animal tissues was believed to bear the closest relation to the presence of blood vessels, and as such to be in complete contrast to the growth of plant tissues, where each part grew independently of all others. Thus the recognition of such instances of 'plantlike' growth in animals, though to us an idea almost without meaning, was of importance at that time.

It was, however, in developing skeletal tissues that the readiest comparison was drawn between the cells of animals and plants, because in both those of cartilage and of the notochord a definite cell wall could be seen. Both Purkinje and Müller had observed cartilage cells in their microscopical studies on embryonic bones, but it was Theodore Schwann who developed this comparison to its full extent.

Schwann's researches in this field are best described by a quotation from his *Microscopische Untersuchungen*, published in 1839 and later translated by the Sydenham Society. His description of the notochord of young fishes and of amphibian larvae tells us something of his methods of research:

> It cannot well be separated entire in recently killed animals, but is best obtained from them in the form of delicate transverse sections. If the animal be placed in water for twenty-four hours or longer after death, and the tail then severed from the body at the point of junction, the *chorda dorsalis* may be entirely pressed out, by gently scraping along its course . . . The interior exactly resembles parenchymatous cellular tissue of plants.

In the cells of both notochord and cartilage he saw nuclei but at first the cell wall seemed the main point of comparison. Schwann speaks of cartilage corpuscles as 'cells in the restricted sense of the word, or as cavities enclosed by a membrane' (*Plate VIII, Fig. H*). This work of Schwann's which surveyed all that then was known of animal cells alongside the description of his own researches is commonly regarded as the inception of the cell-theory, with which we especially associate the two names of Schwann and Schleiden, though until recently the importance of the early observations on animal tissues of Purkinje and his pupils such as Valentin (1810–1883) has not been recognised. The work of this school begins several years before the publication of Schwann's work (Studniˆka, 1927; Florian, 1932). Had Purkinje's contributions to this field not been overshadowed by that of Schwann, the mistaken views of Schleiden might have exercised less influence in subsequent years.

M. J. Schleiden has been described as 'one of the strangest scientific personalities of his age' (Nordenskiold, 1946, p. 392). He was first a barrister, but his lack of success in this career led to an attempt at suicide. On recovery, he resolved to devote himself to science. He became a student once again, and finally attained doctorates in both medicine and philosophy. He then began research into the microscopical anatomy of plants. In 1842 appeared his book, *Principles of Scientific Botany*, a highly individual work, strongly opposed both to the contemporary philosophical and to the systematic treatment of the subject, but insistent that the study of development was the most fruitful approach to the problems of plant structure.

Schleiden's real importance was his power of arousing the interest of younger men in these questions. His influence was responsible for the entry of both Hofmeister and Nägeli into this field; as a microscopist he was largely responsible for inducing the young Carl Zeiss to devote himself to optics (Auerbach, 1926, p. 6). His ascendancy over Theodore Schwann we shall shortly discuss. Schleiden's central idea was of the utmost value, yet in matters of detail he was everywhere wrong. In later years he defended his views with great violence against the younger botanists, who in the first place had owed much to him but had later gradually freed themselves from his errors.

The influence of Schleiden on Schwann was twofold. In the first place he was responsible for establishing in Schwann's mind that it was the cell together with its nucleus which was the real basis of correspondence between the structure of animals and plants. Schwann described how all this began:

> One day, when I was dining with M. Schleiden, this illustrious botanist pointed out to me the important role that the nucleus plays in the development of plant cells. I at once recalled having seen a similar organ in the cells of the notochord, and in the same instant I grasped the extreme importance that my discovery would have if I succeeded in showing that this nucleus plays the same role in the cells of the notochord as does the nucleus of plants in the development of plant cells. [Frédéricq, 1884.]

Here, we have not only the observation of the presence of a nucleus in both types of cell but also a theory of its function and purpose in development. The observation was correct, but the theory, alas, was utterly wrong. And unhappily the young Theodore Schwann unquestioningly adopted the whole of it from his 'illustrious' senior. The history of the cell doctrine for the next fifty years is mainly concerned with the results of this fatal mistake.

What, then, were these false views of the development of cells? Schleiden had really two theories on that subject, both of which were wrong. In one he believed that a daughter cell was formed inside a parent cell, and that the nuclear membrane was transformed into the wall of the new cell (*Plate VIII, Fig. G*). So Schleiden renamed the nucleus the 'cytoblast'. It was no longer present in a mature cell, which might consist only of a wall enclosing a fluid. Schleiden was misled by the ordinary course of development of parenchymatous cells in plants, in which as vacuoles develop within the cell the early condition with a central nucleus gives place to a stage in which the cell sap occupies by far the greater part of the whole volume. Again Schleiden was unfortunate in taking as his typical example of cell formation in a developing plant the events within the embryo-sac after fertilisation. Here is the one instance where cells are formed without laying down walls between them.

Schleiden regarded the contents of the early embryo-sac as a solution of 'gum with intermingled mucous granules'. These granules aggregated, he believed, into larger units equivalent to nucleoli, which in turn grew into nuclei, or cytoblasts. These finally expanded into cells (*Plate VIII, Fig. F*). Such was the conception of the development of cells which became known as 'free cell-formation' and which Schleiden believed to hold good throughout the plant kingdom. Schwann began to apply it to the growth of animal tissues. The doctrine of the 'cytoblastema', the amorphous ground substance out of which cells were produced, was all too readily fitted to the ideas then current of the formation of tissues out of an exudate from blood vessels. Schwann extended the concept of the cytoblastema to include all that we now classify as inter-cellular material. Thus he says (p. 69):

> The chemical and physical properties of the cytoblastema are not the same in all parts. In cartilages it is very consistent, and ranks among the most solid parts of the body; in areolar tissue it is gelatinous; in blood quite fluid.

It was, however, mainly among botanists that Schleiden's views were subsequently discussed. They were vulnerable at two points. In the first place, his account of the formation of cells was only acceptable while no other examples of the process had been described. Furthermore, Schleiden had no understanding of the nature of the primary contents of the cell, about which increasing knowledge in later years revealed a further correspondence between the cells of animals and plants with the recognition of a common living substance within them. Here studies on animal cells led the way, though in the first instance without any reference to cellular structure. In attempts to study the composition of the body of *Hydra*, Abraham Trembley (1700–1784) had teased out portions of its substance in water, and noticed that it seemed to consist of granules held together by 'une matière glaireuse' which was both adhesive and contractile (1744). Little further was learnt about this substance until nearly a century later, when Dujardin (1835) was engaged on studies on the ciliate Protozoa. This author came to reject the conclusions of Ehrenberg (1795–1876), the leading protozoologist of the time, that what we now call the food vacuoles within these organisms were stomachs joined by an intestine. Dujardin concentrated his attention on the substance between these vacuoles. He too found that it was glutinous and contractile, and gave it the name of 'sarcode'. He recognised that a similar material was the basis of the substance of larger creatures, such as flat-worms and annelids. Probably the first observations on the sarcode of single animal cells were those of Valentin in 1836. In a remarkable work on many aspects of the nervous system in both vertebrates and invertebrates he isolated nerve cells and described the appearance of the region between the nucleus and the cell boundary (*Plate VIII, Fig. C*). For this part of the contents of the cell he used the term 'parenchym'.

The word 'protoplasm' was first employed for this purpose by Purkinje in 1839. Theological writers had long used the word 'protoplast' for Adam, the 'first formed'. Purkinje used protoplasm to mean whatever is first produced in the development of the individual plant and animal cell. Among botanists,

Grew's word 'cambium' had already been used in this sense, though without reference to cellular structure. Purkinje's original definition of protoplasm (1840) has been translated by Baker (1949) as follows:

> In plant cells the fluid and solid elements have separated completely . . . In the animal development centre, on the contrary, both are still present in mutual permeation. The correspondence is most clearly marked in the very earliest stages of development—in the plant in the cambium, in the animal in the Protoplasma of the embryo. The elementary particles are then jelly-like spheres or granules . . . With the advance of development the animal and plant structures now diverge from one another; for the former either tarries longer in the embryonic condition or remains stationary in it throughout life, while in the latter on the contrary . . . the separation of the solid and fluid progress more rapidly, and come to light first in cell-formation and then in the formation of vessels.

For Purkinje in 1839, then, the protoplasm of the mature plant cell is represented both by the sap and the cell wall which it has formed. It was only slowly realised that these two categories did not include all the contents of the cell, and that there still is present material which has all the properties of Dujardin's 'sarcode'. One of the most striking aspects of the protoplasm within the plant cell is its powers of movement. The continuous streaming motion which we now term 'cyclosis' was discovered as far back as 1774 by the Abbé Corti in the cells of water plants, such as *Chara* and *Nitella*. His observations were overlooked, and the phenomenon was rediscovered by L. C. Treviranus (1779–1864) in the early years of the nineteenth century (1811). For many years afterwards the movement was usually known as the 'rotation of the cell sap' and at first was regarded as peculiar to water plants alone. It was most easily seen in plant cells with large chloroplasts. Close observation could also reveal the movement of small granules; Robert Brown's great paper of 1833 refers to such a movement in a footnote describing the staminal hair cells of *Tradescantia*:

> A circulation of very minute granular matter is visible to a lens magnifying from 300 to 400 times. This motion of the granular matter is seldom in one uniform circle, but frequently in several apparently independent threads or currents . . . The smallest of the threads or streamlets appear to consist of a single series of particles. The course of these currents seems often in some degree affected by the nucleus, toward which or from which many of them occasionally tend or appear to proceed. They can hardly however be said to be impeded by the nucleus, for they are occasionally observed passing between its surface and that of the cell, a proof that this body does not adhere to both sides of the cavity . . . [1833, p. 712.]

Schleiden himself was familiar with cellular movements of this kind, and in his *Beiträge zur Phytogenesis* (1838) illustrates within the cells of an 'articulated hair of potato' the course of 'a retiform current of mucus upon their walls'.

Schleiden's views on cellular growth were not challenged until a continuous layer of material lining the inside of cells had been demonstrated which consisted not of mucus but of protoplasm. Kützing in 1841 showed the presence of such a layer in some Algae, but regarded it as composed of a precursor of starch, under the mistaken impression that it was changed into that substance under the action of potash. Three years later, K. Nägeli (1817–1891) described this layer in Algae cells as a 'Schleimschicht', and at much the same time H. von Mohl (1805–1872) found a similar structure within the cells of a number of higher plants (*Plate VIII, Fig. J*). Von Mohl gave it the name of 'primordial utricle'. The discovery was first made in material preserved in spirit, where the contents of the cell had contracted away from the wall. Later von Mohl found that strong acids would produce the same result in fresh material; the reversible effect of salt solutions of a suitable strength was not discovered until much later (Pringsheim, 1854; de Vries, 1884). Von Mohl showed that the primordial utricle turned yellow under the action of iodine, and so consisted of nitrogenous material. Two years later he reintroduced the word 'protoplasm' to describe it, apparently independently of Purkinje's previous biological use of the term.

Von Mohl formed the opinion that the first structure that is developed within a young cell is the primordial utricle. He did not regard this view as wholly inconsistent with Schleiden's general theories on the formation of cells, although it demanded some revision of the derivation of a new cell from the nucleus alone. As von Mohl said, Schleiden's view was that

> [the nascent cellular membrane] grows out from the nucleus in such a manner that it is applied upon it like a watch-glass, and the nucleus forms part of the developed cell itself; to me, on the contrary, the cell membrane always appears to surround the nucleus, in the form of a closed vesicle, and in many cases to be at some distance from it. [Trans. Henfrey, 1852, p. 97.]

Von Mohl made a further and still more damaging criticism of Schleiden's theory, according to which

> the cells which were formed in other cells would always be much smaller than the parent cells, and would gradually expand until they filled up the cavity of the parent cells, and their walls came into contact. But as the whole process could not take place in cells which contained granular structures, such as chlorophyll or starch granules . . . without [their] displacement . . ., and yet . . . all these . . . are still present after the division, Schleiden invented an hypothesis to explain the circumstance, namely, that these structures in the cavity of the parent cell were dissolved outside the secondary cell, and formed anew inside it. But as nothing of this process can be observed in nature, it alone suffices to refute the universality of free cell-formation . . . The entire representation proves that Schleiden has never once observed the division of a cell. [Ibid., p. 58.]

By this time several instances had been described of the formation of new cells in plants which could not be fitted into Schleiden's scheme but were clearly examples of a direct division of one whole cell into two daughter cells. Cell multiplication of this kind was most readily observed in filamentous Algae;

examples of early observations of this kind will be given in the following chapter.

Nägeli, who at first had defended Schleiden in the pages of a short-lived journal of which they both were joint editors, turned in 1846 to a comparative study of the production of cells in all groups of plants. He concluded that 'free cell-formation' in the sense of the direct growth of a nucleus into a cell was confined to two instances alone: one of these was found in the development of reproductive cells, the other in the formation of the cells of the endosperm within the embryo-sac of the flowering plant, the original example which Schleiden had assumed to be generally valid. Nägeli, however, concluded that in all other instances in plants new cells are formed by the division of an entire pre-existing cell. This method of division he termed 'parietal cell-formation', as everything within the cell wall is halved to produce two new cells. Nägeli, however, revised Schleiden's views on the origin of nuclei to a much smaller extent. He realised that at least in one instance of parietal cell formation, namely the division of the staminal hair cells of *Tradescantia*, the nuclei of the two new cells were derived by division of the nucleus of the parent cell (*Plate X, Fig. A*). Elsewhere, however, he conceded the nuclei were formed generally as Schleiden had described. Indeed Nägeli regarded both nuclei and cells as 'utricular structures' which could alike be formed in more than one way. Thus further progress in the understanding of the formation of cells waited on a closer grasp of the course of nuclear division. Not until it was realised that the only structures which nuclei can form are other nuclei was it possible exactly to define the separate concepts of cell and nucleus.

This confusion exerted a similar hindrance on the study of the earliest stages of development of the animal embryo. Isolated stages in the cleavage of eggs of several animals had been seen by several early observers, such as Swammerdam (1737), Roffredi (1775; Baker, 1949) (*Plate IX, Fig. A*) and Spallanzani (1780). Prévost and Dumas in 1824 were the first to describe the changes to be seen in living eggs of the frog during segmentation, but these authors had no idea that the changing pattern of furrows on the surface of the egg represented an actual partitioning of the whole into separate units (*Plate IX, Fig. B*). This conception was first reached in 1834 by von Baer (1792–1876), who used the term 'Theilung' to describe the process.

When nuclei were recognised within an embryo in cleavage, however, the difficulties of interpretation increased. Martin Barry (1802–1855), for instance, in 1839 published admirable drawings of cleavage stages of the rabbit in which nuclei are shown. Barry, however, formed the opinion that cells of later stages of development were directly derived by division of these nuclei, and maintained this opinion for the rest of his life (Barry, 1854). Again, Albert Kölliker (1817–1905) described in 1844 the process of cleavage in as difficult an object as the yolky eggs of the cuttlefish *Sepia*; his drawings show in each blastomere a nucleus which is labelled 'Embryonalzelle sammt Kern', for he also then believed that in later stages this structure itself became a cell (*Plate IX, Fig. C*).

Three years before the appearance of Kölliker's work on *Sepia*, however, Robert Remak (1815–1865) gave an example of the real method by which animal cells are formed, in his description of stages of the equal division of embryonic blood corpuscles of the developing chick (*Plate IX, Fig. H*). Remak is one of the greatest names in nineteenth-century embryology, yet racial prejudice denied him a permanent academic career and he was finally diverted into neurological practice. In 1852 while still mainly occupied with fundamental research he reached the conclusion that the division of one cell into two equal daughters was the general method by which cells increased in number during normal growth. In pathological conditions, however, he continued to believe in an endogenous multiplication of cells in which 'the nucleus of the parent cell has no part' (Remak, 1862).

It is understandable that with only the means of preparation of microscopic objects current in the mid-nineteenth century, the study of cellular processes within diseased tissues should have lagged behind the progress of observations on the more accessible tissues in plants and embryonic animals. Indeed, up to 1847, Rudolph Virchow, whose name takes first place in the application of the cell-theory to medicine, still believed in the formation of new cells not even by endogeny but by Schleiden's free cell-formation in a formless blastema.

Rudolph Virchow (1821–1902) was a figure of great importance in the intellectual and social history of nineteenth-century Europe, but not until the appearance in recent years of Dr E. H. Ackerknecht's *Rudolf Virchow, Doctor, Statesman, Anthropologist* (1953) has there been in any language a full-length biographical study of his many-sided career. He began to study medicine at Berlin in 1839 and came under the influence of Johannes Müller. Post-graduate work followed, but was interrupted by the revolution of 1848. While Virchow's youthful radicalism made it advisable for him to leave Berlin, he was fortunate enough to become Professor of Pathological Anatomy at Wurzburg. While still at Berlin he had founded a journal, primarily for pathological researches, still known as Virchow's *Archiv*, and which he continued to edit until his death. Problems of the organisation of tissues were discussed in this journal from its inception. In the first number Virchow stated the views then current among pathologists and histologists on this subject, which had remained unchanged from the time of their adoption from Schwann and from Schleiden. Virchow in 1847 grouped them under three headings:

(1) All organisation progresses through the differentiation of a 'formless blastema'.
(2) Blastema is primarily fluid, and an exudate from vessels.
(3) Differentiation within the blastema results in the formation of cells.

The observational basis of this in pathology is, of course, the phenomenon of inflammation, and the cellular reactions whereby the white corpuscles of the blood in great numbers enter an inflamed area and become macrophages. Schleiden's premature generalisations had led to this process being equated with a false view of the development of the embryo-sac in a flowering plant, under the common name of 'free cell-formation'.

Even in his early days in Berlin Virchow had made observations which seemed inconsistent with all this. While in the eye department of the hospital there, he had seen how the cornea would heal without the appearance of any 'plastic exudation'. At Wurzburg he began to use the microscope intensively to study pathological processes within tissues, and within a few years had come to the same conclusion in this field as his friend, Robert Remak, had done in embryology. Like Remak, he then bracketed free cell-formation with spontaneous generation, and decisively rejected them both.[1] The famous aphorism of Virchow '*omnis cellula e cellula*' appears first in a paper on 'Cellular Pathology' in his own *Archiv* for 1855, mainly concerned, as were most of Virchow's writings at that time, in arguing the wider use of the microscope among pathologists.

In 1856 Virchow was recalled to Berlin, and there began to give the series of lectures under this title, which were first published in one famous volume in 1858. The lectures were addressed to post-graduate students; their object was to expound the bearing of the newer developments in cellular theory upon the understanding of the phenomena of disease and repair. Virchow began with the structure of plant tissues, and revised earlier views on the comparison of the cells of plants and animals. He then proceeded to deal with the structure of the tissues of every organ of the human body, both in health and disease. He greatly extended the definition of the connective tissues and recognised that the cells which we now call 'fibroblasts' were in fact cells, and not nuclei alone. Abnormal proliferation of these cells he considered was the cause of both tubercular lesions and of cancer.

[1] Kölliker in his autobiography (1899, p. 197 ff.) argues at some length that he himself had grasped the concept that cells originate only from pre-existing cells before Virchow embraced it. Again, Schleich (trans. Miall, 1935, p. 157) states that 'Karl Reinhard, a young man of genius, who died at the age of 27, was responsible (according to my father, who knew them both intimately) for the conception that Swann's [sic] cellular theory could be applied to the human organism'.

Virchow's progress at that time was limited by two great gaps in his knowledge. The greater of these was the absence of any understanding of the role of infective agents in disease. Secondly, the elucidation of the processes of nuclear division and of cell multiplication were then still a quarter of a century ahead. Thus it was still possible for him to retain a belief in the endogenous formation of cells, although their production by exogeny had by then been rejected. In malignant tissues, he considered that fresh cells were formed inside others, by a method which he had previously illustrated in 1851. His drawing, however, is clearly of multi-nucleated tumour cells (*Plate VIII, Fig. E*).

Any brief summary of the history of biology almost inevitably is bound to give the impression that with the enunciation of '*omnis cellula e cellula*' and the publication of the *Cellular Pathology* there was a decisive and immediate change in the climate of biological and medical thought. This is not so. In Britain opinions were then much divided on the methods by which cells are formed. As early as 1845, J. D. and H. Goodsir had published their *Anatomical and Pathological Observations*, in which they stated a belief in the growth of a tissue or an organ from the endogenous multiplication of cells within one single cell. This they termed the 'centre of nutrition'. The Goodsirs were strongly influenced by Martin Barry's views on the production of cells by the division of the nucleus of a mother cell. At Edinburgh, in the early 1860's, William Turner, then J. D. Goodsir's demonstrator in anatomy, and afterwards so prominent a figure in that university, was wholly in agreement with the idea of cellular continuity (Turner, 1863), but this school of thought was then opposed by another led by Professor Hughes Bennett, who had been a pupil of Alphonse Donné. Bennett accepted both free cell-formation and spontaneous generation at large (Bennett, 1891). Such views were then still in the majority.

The *Cellular Pathology* reached the English scientific public in a translation of the second German edition, which was published in 1860. It was reviewed in the *British Medical*

Journal with a hostility which recalls the treatment of the *Origin of Species* in the *Edinburgh Review* less than a year previously. Virchow was accused of plagiarism, particularly from J. Goodsir, to whom the *Cellular Pathology* had been dedicated. To the modern reader this review seems a curious mixture of denial of the whole doctrine of '*omnis cellula e cellula*' together with penetrating criticisms of the weaker points of the author's treatment. Thus while the reviewer regards the doctrine of the formless blastema of Schleiden and Schwann as 'consistent with the vast mass of observations since made by histologists, and applicable alike to physiological and pathological processes', he also with justice opposes Virchow's views on endogenous cell-formation, such, for instance, as the origin of pus cells within epithelial cells. The review ends with a few patronising remarks, in the style of Richard Owen, on minutiae where Virchow can be allowed to have observed correctly.

Virchow's task of spreading the practice of microscopy among medical men was undertaken in England by Dr Lionel S. Beale (1828–1906). He had been trained at Oxford by Dr Acland, whose methods of microscopical demonstration we noticed in the previous chapter (Tuckwell, 1907). Later Beale became the successor of R. B. Todd at King's College, London, at first jointly with William Bowman. Beale, in his turn, gave to a medical audience a course of lectures on human histology, which again embodied a theory of the formation and growth of tissues. This series was delivered to the Royal College of Physicians in 1861, and was published in the same year. Beale's views had some elements in common with the cell-theory, but were expressed in an entirely separate terminology. He made a distinction between 'germinal matter' and the 'formed material' which it produced. The latter embraced all products of the cell such as keratin and the intercellular materials of the skeletal and connective tissues. 'Germinal matter', however, has no precise significance; it was sometimes the nucleus alone, sometimes nucleus and cytoplasm together. An *Amoeba*, for instance, consisted entirely of germinal matter. The distinction between germinal material and formed matter was based wholly on no less arbitrary a criterion than the staining reactions of tissues to a solution of carmine, made to a particular formula by which alone Beale is generally remembered at the present day. Insulated by his own ideas and nomenclature, Beale regarded with detachment the various theories of cells and tissues of his day. 'I cannot think', he stated, 'that the cell . . . can be formed from a fluid exudation, but believe with Virchow that in all cases cellular elements must have existed wherever cells are found' (Beale, 1861, p. 61). On the other hand, he thought it probable that the finest living particles of germinal matter that he was able to see were aggregated from still smaller units.

Later in the decade, among the shifting currents of opinion on the nature and formation of cells, one particular stream began to flow with increasing force. The researches on protoplasm in different groups of organisms, such as among Protozoa with Max Schultze (1863) and in slime moulds by De Bary (1864), had led to a re-emphasis of the properties of the living substance common to all organisms. Schultze's studies on the histogenesis of muscle fibres in animal embryos directed further attention to the contents of the cell and away from the cell wall. The word 'protoplasm' was then at hand for the use of the popular lecturer and ready for the enlargement of the common vocabulary. T. H. Huxley (1825–1895) first effected its introduction to a general audience at Edinburgh in November 1868 with his address on the 'Physical Basis of Life'. He was then less concerned with the microscopical appearance and behaviour of protoplasm, than with its philosophical implications. 'All vital action', he inferred, 'may be said to be the result of the molecular forces of the protoplasm which display it' (Huxley, 1869). On such conclusions was built the popular evolutionary materialism of the later nineteenth century, which took the term 'protoplasm' as its symbol, a word once invested only with religious associations. W. S. Gilbert was sure of his laugh when, writing the libretto of the *Mikado* in 1884, he gave Pooh Bah the line 'I can trace my ancestry back to a protoplasmal primordial atomic globule'.

Among biologists also there was a boom in protoplasm in the 1870's. Cleland, in reviewing at that time a number of articles on cellular theory (Cleland, 1873), stated that 'In the present day the protoplasmic element has assumed an enormous importance, casting the nucleus into the shade, while the reign of cell walls has come to an end altogether'. Within ten years, however, the nucleus was to resume its prominence. When to Virchow's aphorism was added the new generalisation *'omnis nuclei e nucleo'* a stable basis for cellular biology was reached at last. But that story belongs to the succeeding chapter.

LITERATURE FOR CHAPTER TWO

ACKERKNECHT, E. H. (1953). Rudolf Virchow, Doctor, Statesman, Anthropologist. Univ. Wisconsin Press.

ARBER, A. (1942). Isis. 34, 7.

AUERBACH, F. (1926). The Zeiss Works and the Carl Zeiss Foundation in Jena . . . Trans. R. Kanthack. London.

BAER, K. E. VON (1834). Arch. f. Anat. Physiol. wiss. Med., p. 481.

BARRY, M. (1839). Philos. Trans. Roy. Soc. London. 128, 301.

BARRY, M. (1854). Monthly Journ. Med. Sci. 18, 518.

BARRY, M. (1854). Monthly Journ. Med. Sci. 19, 10.

BARY, A. DE (1864). Die Mycetozoen (Schleimpilze). 2nd ed. Leipzig.

BAKER, J. (1948). Quart. Journ. Micro. Sci. 89, 103

BAKER, J. (1949). Quart. Journ. Micro. Sci. 90, 87.

BEALE, L. S. (1861). On the Structure of the Simple Tissues of the Human Body. London.

BENNETT, H. (1891). Nature. 43, 14.

BERNHARDT, A. (1834). Symbolae ad ovi mammalium historiam ante praegnationem. Bratislava.

BOSTOCK, J. (1836). An Elementary System of Physiology. 3rd ed. London.

BROWN, R. (1833). Trans. Linn. Soc. 16, 685.

CLELAND, J. (1873). Quart. J. Micro. Soc. N.S. 13, 255.

CORTI, B. (1774). Osservazioni microscopiche sulla Tremella . . . Lucca.

COSTE, P. (1833). Notiz. Geb. Nat. Heilk. (Froriep). 38, col. 241.

DUJARDIN, F. (1835). Ann. Sci. nat. (Zool.) 4, 343.

DUTROCHET, R. H. J. (1824). Récherches anatomiques et physiologiques sur la structure intime des animaux . . . Paris.

EHRENBERG, D. C. G. (1838). Die Infusionsthierchen als vollkommene Organismen . . . Leipzig.

EVELYN, J. (1664). Sylva, or a discourse of Forest Trees . . . London.

FLORIAN, J. (1932). Nature. 130, 634.

FONTANA, F. (1781). Traité sur le Venim de la Vipère. Florence.

FRÉDÉRICQ, L. (1884). Théodore Schwann, Sa vie et ses Travaux. Liège.

GEROULD, J. H. (1922). Sci. Mon. 14, 268.

GOODSIR, J. D., AND GOODSIR, H. (1845). Anatomical and Pathological Observations. Edinburgh.

GREW, N. (1682). The Anatomy of Plants. London.

HENLE, J. (1837). Symbolae ad anatomiam villorem intestinalium . . . Berlin.

HEWSON, W. (1777). Experimental Enquiries. Part the Third. London.

HODGKIN, T., AND LISTER, J. J. (1827). Phil. Mag. 2, 130.

HOOKE, R. (1665). Micrographia. London.

HUXLEY, T. H. (1869). Fortnightly Review. 5, 129.

KÖLLIKER, A. VON (1844). Entwicklungsgeschichte der Cephalopoden. Zürich.

KÖLLIKER, A. VON (1899). Erinnerungen aus meinem Leben. Leipzig.

KÜTZING, F. T. (1841). Linnea, 15, 546.

LAMARCK, J.-B. (1809). Philosophie zoologique . . . Paris.

LEEUWENHOEK, A. VAN (1674). Phil. Trans. Roy. Soc. London. 9, 23.

LEEUWENHOEK, A. VAN (1702). Ibid. **22**, 552.

LINK, D. H. F. (1807). Grundlehren der Anatomie und Physiologie der Pflanzen. Göttingen.

MOHL, H. VON (1837). Flora. 20, pp. 1, 17.

MOHL, H. VON (1845). Ann. Sci. Nat. (Bot.). 3, 71. Trans. A. Henfrey in Taylor's Scientific Memoirs for 1846. 4, 91.

NÄGELI, K. (1844). Zeit. f. wiss. Bot. 1 (1) 34. Trans. A. Henfrey in Roy. Soc. Reports and Papers on Botany. 1846.

NÄGELI, K. (1846). Ibid. 1 (3) 22.

NORDENSKIOLD, E. (1946). The History of Biology. New York.

PRÉVOST, J. L., AND DUMAS, J. B. (1824). Ann. Sci. nat. 2, 100.

PRINGSHEIM, N. (1854). Reprinted in Gesammelte Abhandlungen von N. Pringsheim (1896). Vol. 3, p. 33.

PURKINJE, J. E. (1830). Symbolae ad ovi avium . . . Leipzig.

PURKINJE, J. E. (1840). Uebers. Arb. Veränd. Schles. Ges., vat. Kult. 16, 81.

REICHERT, K. B. (1840). Das Entwickelungsleben im Wirbelthierreich. Berlin.

REMAK, R. (1841). Med. Zeit. 10, 127.

REMAK, R. (1852). Arch. f. Anat. u. Physiol. wiss. Med. p. 47.

REMAK, R. (1862). Quart. J. Micro. Sci. N.S.2, 277.

RICH, A. R. (1926). Bull. Johns Hopk. Hosp. 39, 330.

ROFFREDI, M. D. (1775). Journ. de physique. 5, 197.

ROGET, P. M. (1840). Bridgewater Treatises. V. Animal and Vegetable Physiology, considered with reference to Natural Theology. 2 vols. 3rd ed. London.

SCHLEICH, C. L. (1935). Those Were Good Days. Trans. B. Miall. London.

SCHLEIDEN, M. J. (1838). Arch. f. Anat. Physiol. wiss. Med., p. 137.

SCHLEIDEN, M. J. (1842). Grundzüge der wissenschaftlichen Botanik . . . Leipzig. Trans. E. Lankaster (1849) as Principles of Scientific Botany, London.

SCHULTZ, M. (1863). Das Protoplasma der Rhizopoden und der Pflanzenzellen . . . Leipzig.

SCHWANN, T. (1839). Mikroskopische Untersuchungen . . . Berlin. Trans. H. Smith, (1847). Sydenham Soc., London (with Schleiden's Beiträge zur Phytogenesis, 1838).

SPALLANZANI, L. (1780). Dissertazioni di fisica animale e vegetabile. 2 vols. Modena.

STUDNIĈKA, F. K. (1927). Anat. Anz. 64, 140.

SWAMMERDAM, J. (1737–8). Biblia Naturae . . . Leiden.

TREMBLEY, A. (1744). Mémoires, pour servir à l'histoire d'un genre de polypes d'eau douce . . . Leiden.

TREVIRANUS, G. R. (1805). Biologie oder Philosophie der lebenden Natur . . . Vol. 3. Göttingen.

TREVIRANUS, L. C. (1811). Beyträge zur Pflanzenphysiologie. Göttingen.

TUCKWELL, W. (1907). Reminiscences of Oxford. London.

TURNER, W. (1863). Edinburgh Med. Journ. 8, 873.

VALENTIN, G. (1836). Nov. Act. phys.-med. Acad. Leop. 18, 51.

VAUCHER, J. P. (1803). Histoire des Conferves d'eau douce. Geneva.

VIRCHOW, R. (1847). Archiv. f. path. Anat. Physiol. Klin. Med. 1, 207.

VIRCHOW, R. (1851). Ibid. 3, 196.

VIRCHOW, R. (1855). Ibid. 8, 23.

VIRCHOW, R. (1858). Die Cellular Pathologie. Berlin.

VIRCHOW, R. (1860). Cellular Pathology. Trans. P. Chance. London.

VRIES, H. DE (1884). Jahrb. wiss. Bot. 14, 427.

WEBER, E. H. (1841). Deutsch. Naturf. vers. Bericht., p. 93.

WILSON, J. W. (1944). Isis. 35, 168.

WILSON, J. W. (1947). Ibid. 37, 14.

CHAPTER THREE

Division of the Cell, and of the Nucleus. The Formation of Germ Cells and their Fusion

IN THE previous chapter some account has been given of the history of two false theories of the origin of cells: by exogeny, or free cell-formation within a 'formless blastema'; and by endogeny, or the development of one cell within another. Rather less was then said about early observations on the actual origin of cells by division into two halves of equal size. The process was first observed among the Protista, as might perhaps be expected. Baker (1951, 1953) has recently drawn attention to the fact that Abraham Trembley, after his discovery of fission in several Protozoa in the 1740's, witnessed the same event in the Diatom *Synedra* in 1766 (*Plate VIII, Fig. I*). Twenty years later the division of a Desmid, *Closterium*, was described by the Danish biologist, O. F. Muller; but no further reports of the process appeared until the 1830's, when Ehrenberg began to describe fission in various Protozoa (Baker, 1953). In filamentous Algae, increase in the number of cells by the formation of partitions across them was first seen by Vaucher in 1803 and subsequently also by Dumortier (1832), and also by von Mohl in 1837.

As we have seen in the previous chapter, Nägeli in the later 1840's studied the development of cells in many groups of plants and found that instances of equal division were so numerous that in his survey 'free cell-formation' was restricted merely to a few exceptions. In the terminal cells of the staminal hairs of

Tradescantia, Nägeli (1844) realised that the nuclei of the two daughter cells also were derived from the division of the parent nucleus (*Plate X, Fig. A*). Formation of nuclei by this method, however, Nägeli still regarded as unusual.

Shortly afterwards another worker, W. Hofmeister (1824–1877), studied nuclear division in the same material, yet with a degree of accuracy and detail which rises far above any microscopical observations of the same kind, either at that time or for decades to come (*Plate X, Fig. B*). At that time Hofmeister, like Leeuwenhoek before him, could devote to the microscope merely the hours which could be spared from a life in commerce—'in summer', as his biographer tells, 'from four to six in the morning' (Goebel, 1926, p. 22). He was then in business with his father, a bookseller and publisher in Leipzig, a man of wide interests, particularly in botany and music. At the age of nineteen Hofmeister read Schleiden's *Principles of Scientific Botany* and was so deeply influenced by it that as a self-taught amateur he began to investigate microscopical structure in plants. Within a few years he had advanced sufficiently to publish his first researches. In the later 1840's he wrote papers on such fundamental subjects as fertilisation, the development of the embryo in flowering plants, and on cell-division. He was awarded an honorary doctorate by the University of Rostock in 1851, but an academic post was not offered to him until twelve years later, when he became Professor of Botany at Heidelberg. In 1872 he succeeded to von Mohl's chair at Tübingen.

In his earliest work on cell-division he saw that both in the pollen-mother and staminal hair cells of *Tradescantia* the nuclear membrane dissolved before division of the cell, but that its contents were then still visible. Round this material he thought there collected a mass of 'granular mucilage'. Before the cell divided, the whole parted into two masses; a membrane formed round each, and so two daughter nuclei were produced (*Plate X, Figs. B and D*). It may be said that no comparable observations consciously directed to the understanding of the division of nuclei were made for another thirty years. To Hofmeister's observations on fertilisation we shall later return.

Some of the earliest observations on the division of animal cells were made on cleavage stages in eggs; but, as was said in the previous chapter, the nature of the process was not at first understood. Even after the idea of the multiplication of cells by division was current, it was still possible to describe the division of an egg into blastomeres and not to be aware that cells had thereby been formed; or even if that were admitted, to deny that the new cells would form part of the later embryo.

There were, however, workers even at that time who saw something of the stages of nuclear division in animal cells; Remak, for instance, in 1841 described and correctly interpreted pairs of blood corpuscles which were seen in the blood of a late chick embryo. They were connected by a stalk in which a strand ran between their nuclei. Baker (1955), in the fifth of his valuable series of papers on the history of each aspect of the cell theory, has discussed the early observations at this epoch on the formation of nuclei by division.

Within this period a few observers noticed a radiating appearance within egg cells during cleavage, which we now know to belong to the earliest stages of nuclear division. In 1847 Derbès saw the asters within sea-urchin eggs (*Plate IX, Fig. F*), as also did Reichert within the spermatocytes of the Nematode *Ascaris* (*Plate IX, Fig. G*), an animal which was later to play so important a part in nuclear cytology. In 1852 Krohn saw the whole astral configuration in the egg of the ascidian *Phallusia* and described the double system of diverging rays.

Reichert (1811–1883) had concluded, on the basis of these studies on Nematode spermatocytes, that when a cell prepares to divide the nucleus is dissolved. Within each daughter cell he believed a new nucleus is again formed. Meanwhile, however, Remak had continued his researches on the multiplication of cells in the embryo, and by 1852 had decided that the nuclear material did persist from one cell generation to another. In 1858 he came to the conclusion that the division of both cell and nucleus was 'centrifugal', nucleolus, nucleus, cytoplasm and cell membrane in turn dividing by simple constriction. In this paper he returned to the study of the blood cells of the chick, which he then examined after treatment with dilute potassium bichromate. From a five-day chick he illustrated erthyoblasts in division at stages which we can recognise as anaphase and telophase (*Plate IX, Fig. H*).

A few years later the French protozoologist E. G. Balbiani (1825–1899) was studying conjugation in some ciliate Protozoa. The plates which illustrate his paper in 1861 on this subject show with remarkable accuracy and fine detail much of the complex cycle of events which then occurs within these organisms; Balbiani was among the very first to use a fixative and a staining reagent to give a selective effect; he used acetic acid followed by a very weak solution of carmine, whereas at that time this substance was usually used in so concentrated a form that only a general coloration resulted. Unfortunately Balbiani totally misunderstood his own exquisite preparations. For him, as for Ehrenberg, the Ciliata were 'complete animals', with intestines and reproductive organs comparable to those of macroscopic animals. For him the macronucleus was the ovary and the micronucleus the 'testicule'. In his figures of the latter we can see several phases of its mitotic division; both the crescentic prophase and the later stages when the spindle has formed (*Plate X, Fig. G*). The spindle fibres and the chromosomes in their metaphase were all too readily identified as a bundle of spermatozoa.

Thanks in large part to this sad error, the discovery of the full sequence of changes during nuclear division was postponed to the following decade. Balbiani himself was then among those who made observations on the dividing nucleus, but only much later (1892) seems to have realised that he had already figured a dividing nucleus more than thirty years before. It was not until 1876 that Bütschli showed that the structures within a conjugating pair of ciliate Protozoa which Balbiani had figured were concerned with the division of the micronucleus.

The great majority of papers in the 1870's in which nuclear division was studied in animal cells were concerned with early stages of cleavage of the egg. Side by side with the study of the division of the nucleus of the fertilised egg there proceeded the investigation of the events which led to its formation, of the fusion of the pronuclei of the egg and sperm. For this reason it is here necessary to retrace our steps at this point and describe how the understanding of the nature of fertilisation had by then reached this point.

In plants there were microscopical observations on the events during fertilisation, even before final conviction on plant sexuality had been reached. Here, J. B. Amici (1784–1860) is the most important figure; his researches in this field went on side by side with his technical investigations in optics. In 1823 with his reflecting microscope he saw an isolated pollen grain put out a pollen-tube. Seven years later he followed the process *in situ* and observed an ovary into which a pollen-tube had entered. Robert Brown, in 1833, found these tubes in the ovaries of pollinated orchids, but expressed some doubts on their origin. In 1837 Schleiden entered this field; he confirmed the origin of the pollen-tube, but asserted that it grew into the embryo-sac and itself there formed the embryo. Thus, according to him, the pollen grain was the female element of a flowering plant. The usual views on the sexes of plants, however, survived this assertion, and few botanists accepted Schleiden's conclusion. Amici opposed it, in the face of considerable abuse from Schleiden. By 1846 Amici had proof of the contrary view and was able to show that in orchids an egg-cell, already present in the embryo-sac of the ovule before the arrival of the pollen-tube, was afterwards stimulated to develop into an embryo (*Plate XI, Fig. A*). Hofmeister in 1849 confirmed this sequence of events in a variety of flowering plants, but, like Amici, had then no notion of the actual nature of the stimulus which the pollen-tube exerted.

During the next decade fundamental observations were made on the nature of fertilisation in certain Algae. In *Fucus*, the common brown seaweed, the large egg-cells which are set free into the water were familiar objects, and in 1854 Thuret saw an egg surrounded by much smaller motile bodies, the ciliated spermatozoids. Some of these were attached to the wall of the egg-cell. From that time Thuret's drawing of this event has provided the standard illustration of the process in this alga. He collected spermatozoids and ripe egg-cells separately, performed an artificial fertilisation and even succeeded in hybridising different species of the genus. In 1855 N. Pringsheim observed fertilisation in *Vaucheria*, a fresh-water alga, and in the following year saw the entrance of a spermatozoid into the egg-cell of *Oedogonium*.

By this time much the same level had been reached in the study of fertilisation in animals. In 1824 Prévost and Dumas showed in the frog that it was the spermatozoon and not the seminal plasma that was the essential element in fertilisation, thus bringing to a final conclusion the experiment which Spallanzani had attempted in 1786. Nearly thirty years after the work of the two French authors, George Newport (1852) resumed the study of the phenomenon in this animal. He sent a short note to the Royal Society in which were these words:

> I have ascertained that the spermatozoa of the Frog are not only brought into contact with the surface of the egg in fecundation, as already known, but that some of these bodies penetrate into the thick gelatinous envelopes as stated by Prévost and Dumas, and further I have found that in those eggs which are completely fecundated, some spermatozoa have arrived at and become partially embedded in the internal envelope which encloses the yolk, although I have not yet been able to detect any within the yolk itself: nor have I obtained any evidence of the existence of an orifice or natural perforation in the external envelopes through which these bodies might enter.

Within a mammalian egg, however, spermatozoa had already been observed nearly ten years before. In 1843 Martin Barry published a short note on fertilised eggs of the rabbit taken from the Fallopian tube, but this observation was disbelieved for many years.

In the frog's egg, Newport was able to show that the point at which the sperm entered the egg determined the plane of the first cleavage furrow, yet the fact that fecundation is the work of one single spermatozoon did not emerge from his observations. Indeed Darwin, in his theory of Pangenesis (1868), laid some stress on the apparent need for several male elements, which he cited from Newport. It was not until 1879 that Hermann Fol (1845–1892) was able to see that in the starfish egg it is only one spermatozoon which enters (*Plate XI, Fig. D*).

On the events subsequent to fertilisation, one important observation was made at the mid-century, although its significance remained obscure for many years. Nicholas Warneck, a Russian biologist, described the changing appearances of the eggs of fresh-water Gastropods after laying (*Plate IX, Fig. D*). On these observations, E. L. Mark commented in 1881 that 'as regards what may be observed on the living egg [they] . . . leave very little room for additions'. Warneck (1850) noticed that within the freshly laid egg two rounded bodies could be made out, but a little later only one was to be seen.

It was not until late in the 1870's that the meaning of Warneck's observations was understood. During this decade the intensity of research on fertilisation and on nuclear division very greatly increased. After 1875 the rate at which papers in these fields were published shows a marked rise (Hughes, 1957). At that time it was not uncommon for the leading cytologists, most of whom worked in German laboratories, to publish up to seven papers a year. This may well be the first branch of biology which attracted research on what we now regard as a modern scale.

From this extensive fabric of investigation we can but pick out a few of the more prominent threads. In 1873, O. Bütschli (1848–1920) saw that there were two nuclei within a fertilised Nematode egg about to begin division. This was no more than a confirmation of Warneck's observations, and its significance was then still no less obscure. In the following year, however, Auerbach observed that these nuclei fused, and recognised this process as one of conjugation. Shortly afterwards, in 1876, Oscar Hertwig (1849–1922) realised that in the fertilised sea-urchin egg one of these two nuclei was derived from the spermatozoon.

These researches on fertilisation in the eggs of animals went on side by side with studies on the division of the nucleus after the completion of this event. Often a paper described observations in both topics, to each of which it may represent a contribution of great importance. Among the many studies on the animal egg at that time were further observations on the system of fibres and rays which was later termed the 'achromatic figure', and which had first been noticed within the dividing sea-urchin egg nearly thirty years before (p. 57, above). In the fertilised egg of *Geryonia*, Hermann Fol in 1873 described the whole system of the spindle and the astral rays, and compared it to the lines of force between a pair of opposite magnetic poles. This analogy, though a misleading one, was later to contribute to a good deal of speculation on the mode of action of the whole structure. Most observers at that time, however, were concerned with the source of the material from which it was formed. Auerbach in 1874, on the basis of studies on eggs, both of the Nematode *Ascaris* and of the Echinoid *Strongylocentrotus*, thought that the whole structure was formed from nuclear sap, which he considered dissolved the remainder of the nucleus. So he gave the spindle and asters the name of 'karyolytic figure'. Bütschli (1875), however, thought that in the eggs of free-living Nematodes and of snails the spindle was formed from the nucleoli, but in a second paper in the same year then came to regard the spindle as the product of the whole nucleus. This was also the view expressed by Oscar Hertwig (1876) in the first of his papers on the sea-urchin egg, to the observations of fertilisation in which we have already referred. In this year van Beneden (1845–1910) discovered the 'corpuscule polaire', or central body in the cells of the obscure Mesozoan Dicyemid parasites of Cephalopods. It was only in the following decade, however, that further studies by the same author led to the derivation of the aster from the central body and the recognition of its importance as a permanent organ of the cell (van Beneden, 1883; van Beneden and Neyt, 1887). Meanwhile in 1875 the spindle of the dividing plant cell had been discovered. Eduard Strasburger (1844–1912) began his researches in plant cytology with a study of cell division in the embryo of a conifer. His figures clearly show a fibrous spindle in these cells at several stages of division. These and other observations in the same field were described in the first edition of his *Zellbildung und Zelltheilung* which appeared in 1875.

Even at that time there was no general agreement on the question of what became of the original nucleus when a cell divided. Auerbach's ideas on the dissolution of the nucleus by the 'karyolytic figure' represented little advance on those which Reichert had stated in 1847. Walter Flemming (1843–1915), who towards the end of the 1870's became the leading figure in the study of cell division in animals, earlier in the decade was also of the opinion that the process began with the breakdown of the original nucleus. In 1874, within the eggs of fresh-water Lamellibranchs he saw stages which showed either an aster or a nucleus, but never both together. Even when he had observed a metaphase plate in the same material he still maintained this opinion (Flemming, 1875). In 1873, however, Fol, in his paper on the egg of *Geryonia*, had seen that at a stage when the nuclear membrane has just disappeared treatment of the cell with acetic acid would bring back into view what still remained of the nuclear contents.

It was not until cytologists had come to realise how greatly cells differ from species to species in the clarity of their division figures and in the ease with which the details can be observed that the full sequence of changes in nuclear division was revealed in favourable material. Schneider in 1873 had published figures of the later stages in the division of the eggs of the Platyhelminth *Mesostomum*, but the choice of the most suitable cells for studies of this kind came only in the second half of this decade. It is at this time that the whole field shows such a marked increase in activity. In 1875 van Beneden gave an account of nuclear division in the cells of the rabbit blastoderm. He described how from the nucleus there came an 'essence nucleaire' which formed the equatorial plate. This then divided into two discs. In the following year Balbiani (1876) described his observations on the process as seen in the ovarian epithelium of the grasshopper *Stenobothrus*, a species which, as we now know, is particularly suitable for such studies. From the original nucleus were formed a number of 'batonnets étroits', each of which had the form of a row of globules. The 'batonnets' became arranged in a bundle; each, Balbiani said, divided in the middle, and thus the secondary bundles were formed, each of which, as the cell divided, became a daughter nucleus by fusion into a homogeneous mass. By that time a number of other observers had described the formation of the daughter nuclei from a paired mass of threads or granules.

Balbiani's account contains only one major error, namely the division of the 'batonnets' in the middle. Within a few years, however, more accurate observations were made on this point by Flemming (1879), who by that time had chosen to study embryonic cells of *Salamandra*, in which the finer details of the division process could be readily observed. Here Flemming saw clearly that the division of the nuclear threads was a longitudinal one.

In the next year, 1879, a new method of studying the dividing cell was introduced. In that year three observers published accounts of what they had seen of the process in living cells under the microscope, under conditions in which the course of division continued during the period of observation. Strasburger (1879) was able to follow the division of living staminal hair cells of *Tradescantia* (*Plate X, Fig. C*), while thin slices of embryonic amphibian cartilage were chosen for this purpose by Flemming himself, and also by Schleicher; these two observers published their results independently in one volume of the same journal. These investigations confirmed that the various stages of the division process did in fact succeed each other in the order which had already seemed most probable, and gave a striking illustration of the fundamental fact that the process, in all but minor details, is common to both the plant and animal kingdom. A name was obviously needed for so general a phenomenon; Schleicher chose 'Karyokinesis' (i.e. nuclear movement), a word still used in continental Europe, and possibly a better one than Flemming's later term 'Mitosis' which relates only to the behaviour of the nuclear threads.

So between 1875 and the end of the decade general agreement was reached, that not only are cells in both animals and plants always formed by equal division, but that division of the nucleus precedes that of the cell. Comparison of the first and third editions of Strasburger's *Zellbildung und Zelltheilung* suggests that we may assign to these years the full acceptance of this conclusion. Even in the first edition in 1875 the author still maintained that within the embryo-sac of the higher plants the endosperm was formed by Schleiden's method of free cell-formation. By 1880, however, when the third edition was published, Strasburger had realised that here was a special form of the normal process of karyokinesis, whereby a large number of nuclei are rapidly produced, but where the formation of cell-walls between them follows at a more leisurely pace (*Plate XI, Fig. B*). Generally in plants daughter cells become separated by the laying down between them of a plate of cellulose, which is developed within a fibrillar structure, the phragmoplast. This succeeds the mitotic spindle, and has a similar appearance.

The term 'free cell-formation' was somewhat misleadingly retained for the process of cell-division within the embryo-sac. Strasburger's later illustrations of the process show many nuclei in simultaneous division; in recent years a Polish botanist, Dr Bajer, has found in the dividing endospermal nuclei an admirable source for the study of mitotic figures in living material.

Early in the 1880's a synthesis of the cytological researches of the previous years was published by Walther Flemming under the title of *Zellsubstanz, Kern und Zelltheilung* (1882). Flemming's book differed from Strasburger's in several respects. In the first place the emphasis is naturally more on animal cytology. He was, moreover, much concerned with the composition of the intermitotic cell; his views on the structure of protoplasm we shall discuss in a later chapter. The most important section of the book, however, deals with his own definitive observations on mitotic division. Here he was able to show what could be accomplished when, with the aid of his own accurate cytological techniques, suitable material was studied by means of the homogeneous oil-immersion lens (*Plate X, Fig. F*).

Flemming distinguishes nine phases in the process of nuclear division, of which the early steps within the mother nucleus are repeated in reverse towards the end of the cycle within each of the daughters. The earliest phases are perhaps the most difficult of all to study, and here Flemming had much to contribute. His account of what we now term 'prophase' was derived from amphibian nuclei, in which the form and behaviour of the chromatin threads can readily be made out. One major point, however, on which we now differ from Flemming is that in the first phase he thought the fine chromatin threads within the nucleus formed a continuous ball, the spireme, which was later subdivided into separate lengths. We now believe that these threads are separate from the first. When the nuclear membrane has disappeared each thread is V-shaped, and all are radially arranged around the spindle. The whole configuration, Flemming considered, was held together by an attractive centre, situated at first in the mid-point of the spindle. This then divided into two and its halves moved to the poles of the spindle, carrying with them to each pole a group of halved chromatin loops. From each set a daughter nucleus was reconstituted. In the following years Flemming's account of the process of mitotic division was amplified and revised, both in regard to the development and form of the achromatic figure, and also to the behaviour of the loops of chromatin, for which in 1888 Waldeyer introduced the term 'chromosome'.

Although Flemming regarded their division as a longitudinal one, both in animals and plants, Strasburger, in the third edition of his book, still maintained that the chromosomes divided transversely. A few years later, however, he revised his opinion, for by then there had appeared several further accounts of the longitudinal splitting of chromosomes in a number of species, both plant and animal. In 1883 Guignard reported that he had observed this method of division in both pollen mother-cells and in the embryo-sac of a number of flowering plants, both mono- and dicotyledons. There were attempts to enumerate chromosomes as early as 1878 by Selenka in the eggs of the Echinoderm *Toxopneustes*. Irregular numbers were counted, though it is understandable how no constancy was then detected among the numerous small ones of Echinoderms. However, in *Salamandra* both Flemming and Carl Rabl found that there were always twenty-four; Guignard observed fixed numbers in *Lilium* and in other plants with large chromosomes.

We owe the concept of their persistence during interphase, though in a form in which they can no longer be separately discerned, in the first place to Rabl (1885). The early daughter-nucleus in *Salamandra* is kidney-shaped; centred on its hilus in telophase, the chromosomes become radially arranged. Again the same pattern is seen in early prophase. So the suggestion arose that during the intervening period the chromosomes, although not then separately recognisable, kept their relative positions, yet in some masked form. It was in the Nematode *Ascaris megalocephala*, however, that the individuality of the single chromosome was first demonstrated. Soon after van Beneden began his researches on fertilisation in this species it became clear that two varieties could be distinguished, one with only two and the other with four chromosomes in the fertilised egg. In the former variety, named *univalens* by van Beneden, the opportunities for the study of the behaviour of the chromosomes in fertilisation and division were extremely favourable. The studies of van Beneden and later of Boveri on both varieties of this organism will require notice under several headings.

During the course of reconstruction in telophase in the nuclei of early cleavage stages the free ends of the chromosomes project in such a way that as the nuclear membrane forms round them finger-shaped lobes are produced (*Plate XI, Fig. F*). Boveri was able to show them in *A. megalocephala univalens*. Not only did the arrangement of the chromosomes from telophase to the succeeding prophase remain unchanged but it was identical in sister nuclei.

By this time much had been learnt concerning the complex events by which the egg-cell is prepared for fertilisation and how, moreover, this event is finally accomplished. The nature of the bodies which are extruded from the egg and were first called 'globules polaires' by Robin in 1862 was for long wholly obscure, as was also the cause of their expulsion. They are readily seen in the eggs of fresh-water Mollusca, and were first observed as early as 1824 by Carus. In 1850 Warneck gave clear figures of their origin in *Limax,* and about that time Loven (1848) suggested that the polar body was formed by extrusion of the nucleolus of the egg-cell. However, by the time that something of the general nature of nuclear division was understood, it became clear that they arose by a process of unequal cell division. In 1877 both O. Hertwig and Giard reached this conclusion with respect to the Echinoderm egg (*Plate XI, Fig. C*).

In 1881 E. L. Mark reviewed the literature on this subject at that time, and came clearly to the conclusion that polar bodies were cells. On the question why the egg should undergo this peculiar form of division before the union of the male and female nuclei he could only offer the suggestion that they represented a phylogenetic rudiment of some earlier form of asexual reproduction. Some years previously, however, Bütschli (1876) had been able to show experimentally that in Nematodes fertilisation and the formation of polar bodies are causally related. He succeeded in isolating and rearing females of *Rhabditis,* and found that in the absence of the former event the second did not occur.

It was again in *Ascaris megalocephala* that the whole chain of events in the maturation of the germ cells and fertilisation was finally disclosed, though researches in several laboratories extending over a number of years were necessary before the whole process was clear. By 1883 van Beneden understood the essential nature of fertilisation; in *Ascaris* it is not until the male and female pronuclei have formed their chromosomes that the respective nuclear membranes break down. Each set of chromosomes then moves to the equatorial plate (*Plate XI, Fig. E*). Van Beneden could thus see that in the *univalens* variety each parent contributes a single chromosome to the pair possessed by the zygote. In *bivalens* the corresponding numbers are two chromosomes in each of the pronuclei, and four in the zygote. He further realised that in giving off the polar bodies the number of chromosomes within the egg nucleus is reduced from the double to the single number. However, van Beneden did not then correctly describe how this reduction was brought about; he was right in regarding the nuclear division of the ripe egg as a special type, but thought that its peculiarity lay in that the chromosomes remain undivided. Thus he thought that of the four chromosomes of *bivalens* two entered the first polar body and two remained to form the female pronucleus.

However, Boveri in 1887–8 and O. Hertwig in 1890 separately discovered the real nature of reduction division in the egg; Hertwig drew attention to its essential identity with the corresponding process in the maturation of the sperm.

In the final 'maturation' divisions which lead to the formation of ripe gametes in both sexes each chromosome in the mother-cell divides once, but the cell itself divides twice. Thus in the spermatogenesis of *Ascaris megalocephala bivalens* four chromosomes split longitudinally. Of the eight daughter chromosomes which are thus formed, two are distributed to each of four developing spermatozoa. In a general review on fertilisation published in 1892 T. Boveri (1862–1915) showed that exactly the same scheme holds for the maturation of the ovum if the first polar body divides after it has been given off. Thus in the course of oogenesis in this organism, into each of four cells, namely three polar bodies and the ovum, go two of the eight daughter chromosomes of the egg mother-cell.

As the course of events in the fertilisation of the egg in other species was investigated it became clear that the details of the whole process were by no means everywhere the same. In the Echinoderm egg, for instance, as Wilson (1895) showed for *Toxopneustes*, both polar bodies have been formed before the egg is shed into the water to meet the spermatozoa. When the sperm head has penetrated the egg it develops into the male pronucleus, which immediately fuses with the female nucleus. Both nuclei at the time of fusion are still in an interphase condition, where their chromosomes are individually unrecognisable. It is now known that the sea-urchin and the Nematode

Ascaris represent the opposite ends of a series with respect to the course of events in the fertilisation of the egg, though most animals stand nearer to the *Ascaris* end of this line. Thus in the type of fertilisation which is represented by this organism the history of the chromosomes contributed by each parent can more readily be made out, while in the sea-urchin egg, on the other hand, the behaviour of the asters is less intricate, because those concerned in the formation of the polar bodies have disappeared by the time when the spermatozoon has introduced a further astral figure.

Since the work of van Beneden and Neyt, and also of Boveri in 1887, it has been known that the centrosome was a permanent cell organ which remained within the cell during the resting period, and divided into two just before the next mitosis began. In early prophase, the daughter centrosomes take up their positions at opposite ends of the nucleus, and from them is built up the spindle, together with the astral figures (*Plate XII, Fig. A*). From this concept of the centrosome as a permanent cell organ, Rabl suggested that in fertilisation conjugation of the pronuclei should be accompanied by an apposition of paired paternal and maternal centrosomes at each pole of the zygote nucleus. This configuration Hermann Fol proceeded to discover, and in 1891 described its aspect in the sea-urchin egg under the arresting title of the 'quadrille of the centres' (*Plate XII, Fig. C*). This paper attained considerable notoriety, and a number of other observers in the following years showed division figures with two centrosomes at each pole of the spindle in a number of animals and plants. Among such were Guignard's description of the maturation divisions in the Lily (1891), though, as we now know, centrosomes are unknown in the cells of the higher plants. Guignard's paper, however, is of importance as one of the first demonstrations in plants of reduction division.

The centrosomes of the sea-urchin egg during the course of fertilisation were re-investigated by E. B. Wilson (1856–1939) in 1895; he dismissed the quadrille of the centres as but a 'contribution to biological mythology'. Wilson found that the asters of the female nucleus at the formation of the second polar body disappeared without trace, and that the ripe unfertilised egg is without a centrosome until the entry of the middle-piece of the spermatozoon. This body, itself formed during spermatogenesis by a centrosome, provides the division centres for the mitotic figure of the zygote nucleus, and, as Boveri suggested in 1887, in turn for all subsequent cell-divisions of the same individual. Boveri brought forward a variety of evidence which suggested that it is the sperm aster rather than the male pronucleus which provides the essential stimulus in fertilization. Whether, however, it is true that in all eggs the original sperm aster and its descendant centrosomes exerts an influence which spreads through all the later cell generations depends on how far we can accept the thesis that a central body never originates *de novo.* On this question final agreement has never yet been reached. In the higher plants alone the absence of centrosomes is sufficient to show that Boveri's theories are not universally valid among all cellular organisms.

The formation of the gametes and their fusion were studied at much the same time in both plants and animals. As early as 1877 Strasburger published a general paper on fertilisation and the early stages of embryogenesis. In conifers and a number of flowering plants the two nuclei of the pollen-tube were described, as was the fusion of one of them with the egg nucleus. In the Orchid *Monotropa* he observed the fusion of the two nuclei of the embryo-sac. Among lower plants he saw a conjugation of nuclei also in *Spirogyra,* but then thought that here the fusion nucleus was dissolved. Two years later F. Schmitz (1879), in the same alga, recognised the true position which this nucleus occupies in the whole life-cycle.

The full sequence of events in the fertilisation of the higher plants was not revealed until the end of the century, when S. Navashin in 1898 discovered in *Lilium* the double process of nuclear fusion, in which not only does one male nucleus unite with that of the egg as in all sexually reproducing organisms, but also a second nucleus from the pollen-tube fuses with two other nuclei at the centre of the embryo-sac to form the primary nucleus of the endosperm. In the following year Guignard (1899) described the same events in more detail.

Two years previously this author had been able to confirm an even more important discovery. In 1888 Strasburger began to publish a series of 'Histologische Beiträge', in the first of which he showed that at the time when the gametes in flowering plants are formed the number of chromosomes is halved during the nuclear divisions which lead to the formation of both the pollen-grain and the embryo-sac. Guignard's papers on reduction division appeared soon afterwards in 1889 and 1891. They initiated a debate on the precise manner in which the halved numbers of chromosomes was attained.

The significance of the process in the life-cycle of the plant was made clear in a paper published by E. Overton in 1893, only four pages in length. He showed that the halved number of chromosomes was characteristic not merely of the gametes but also of the whole sexual generation, the gametophyte, which in the flowering plants has been reduced almost to extinction, but which, however, in the more primitive members of the Gymnosperms is less vestigial. Overton found that in the Cycad *Ceratozamia* the nuclei of the endosperm, which here represents the female prothallus, contain but eight chromosomes, while sixteen were to be found throughout the sporophyte generation. He extended his researches to the mosses and ferns to see whether the events within the spore mother-cells of the sporophyte generation corresponded with those in the formation of the pollen-grains in the higher plants. Although Overton was not able to make precise counts of the chromosomes in those Cryptogams which he chose for study, he inferred that the expected reduction took place. Furthermore, he stated that the

> details of karyokinesis [sic] in the spore mother-cells of these plants correspond exactly to those seen in the mother-cells of the pollen. There is the same protraction of the first phases of division, the same thick and excessively short chromosomes, and the same early longitudinal division.

In the following year J. B. Farmer counted the chromosomes in both generations of the Liverwort *Pallavicinia*, and found that in the formation of the spores the number of chromosomes was reduced from eight to four.

Up to this time the terminology of the maturation divisions had been that introduced by Flemming in 1887. The first phase, clearly different from a mitotic division in somatic cells, Flemming called 'heterotypic'. For Flemming the second phase of maturation was an example of the usual process of the division of a nucleus and was called 'homotypic'. With the recognition that the number of chromosomes was halved during maturation, another term for the whole process was needed. It was Farmer who in a paper with J. E. Moore in 1905 proposed a name for the full cycle, based on the Greek verb *μειοῦν*, to lessen. Their form of the term was spelt 'maiosis', which was altered by later writers to 'meiosis', a word already in existence, with the meaning of 'understatement'. The universal adoption by cytologists of this term in a biological sense has not yet been acknowledged by the Oxford Dictionary.

The recognition that meiotic division is found in all sexually reproducing organisms whose chromosomes are visible, both among animals and plants, was based on surveys ranging widely throughout both kingdoms. Thus was established a further branch of comparative anatomy. By now, study of the number and behaviour of the chromosomes in different organisms has taken its place as one of the major branches of evolutionary biology (White, 1945). Detailed analysis of this kind in related organisms has disclosed fresh aspects of the problems of the definition of species and has indicated something of the probable course of evolution of such specialisations as parthenogenesis. The subject of organic evolution, however, has always been integral with the study of heredity and of variation, and upon these fields the growth of cytology has exercised the profoundest influence. In the next chapter will be given some account of this development.

LITERATURE FOR CHAPTER THREE

AMICI, J. B. (1823). Quart. J. Sci. 16, 388.

AMICI, J. B. (1830). Ann. Sci. Nat. (Bot.) 21, 329.

AMICI, J. B. (1846). Atti. Scienz. Ital., p. 542.

AMICI, J. B. (1847). Ann. Sci. Nat. (Bot.). 7, 193.

AUERBACH, L. (1874). Organologische Studien. Breslau.

BAJER, A. (1951). Acta. Soc. Bot. Poloniae. 21, 1.

BAKER, J. R. (1951). Isis. 42, 285.

BAKER, J. R. (1953). Quart. J. Micro. Sci. 94, 407.

BAKER, J. R. (1955). Quart. J. Micro. Sci. 96, 449.

BALBIANI, E. G. (1861). Journ. de Physiol. 4, pp. 102, 431, 465.

BALBIANI, E. G. (1876). Comptes Rend. Acad. Sci. Paris. 83, 831.

BALBIANI, E. G. (1892). Rec. Zool. Suisse. 5, 1.

BARRY, M. (1843). Philos. Trans. Roy. Soc. London, p. 33.

BENEDEN, E. VAN (1876). Bull. de l'Acad. Roy. de Belge. 2nd Ser. 41, 1160.

BENEDEN, E. VAN (1883). Arch. de Biol. 4, 265.

BENEDEN, E. VAN, AND NEYT, A. (1887). Bull. de l'Acad. Roy. de Belge. Ser. 3, 14, 215.

BOVERI, T. (1887). Anat. Anz. 2, 688.

BOVERI, T. (1888). Jen. Zeit. 22, 685.

BOVERI, T. (1892). Anat. Hefte Abt II. 1, 386.

BROWN, R. (1833). Trans. Linn. Soc. 16, 685.

BÜTSCHLI, O. (1873). Nova acta phys.-med. Acad. Leop. 36, No. 5.

BÜTSCHLI, O. (1875). Zeit. f. wiss. Zool. 25, 201.

BÜTSCHLI, O. (1876). Abhandl. Gesell. Naturf. Senkenberg. 10, 213.

CARUS, G. G. (1824). Von den äusseren Lebensbedingungen der Tiere . . . Leipzig.

DARWIN, C. (1868). Animals and Plants under Domestication. 2 vols. London.

DERBÈS, M. (1847). Ann. sci. Nat. (Zool.) Ser. III, 8, 80.

DUMORTIER, B. C. (1832). Nova Acta phys.-med. Acad. Leop. 16, p. 217.

FARMER, J. B. (1894). Ann. Bot. 8, 35.

FARMER, J. B., AND MOORE, J. E. (1905). Quart. J. Micro. Sci. 48, 489.

FLEMMING, W. (1874). Arch. f. mikro. Anat. 10, 257.

FLEMMING, W. (1875). Sitzber. d. k. Akad. wiss. Wien. Math. nat. 71, 81.

FLEMMING W. (1878). Schriften des naturw. Vereins für Schlesw-Holstein.

FLEMMING, W. (1879). Arch. f. mikro. Anat. 16, 302.

FLEMMING, W. (1882). Zellsubstanz, Kern, und Zelltheilung. Leipzig.

FLEMMING, W. (1887). Arch. f. mikro. Anat. 29, 389.

FOL, H. (1873). Jen. Zeit. 7, 471.

FOL, H. (1879). Mém. Soc. Phys. et Nat. Genève. 26, 89.

FOL, H. (1891). Anat. Anz. 6, 266.

GIARD, A. (1877). Comptes rend. Acad. Sci. Paris. 84, 720.

GOEBEL, K. VON (1926). Wilhelm Hofmeister. Trans. H. M. Bower, London. Roy. Soc.

GUIGNARD, L. (1883). Comptes rend. Acad. Sci. Paris. 97, 646.

GUIGNARD, L. (1889). Revue Gén. Bot. 1, pp. 11, 63, 136, 175.

GUIGNARD, L. (1891). Ann. des Sci. Bot. 7 ser. 14, 165.

GUIGNARD, L. (1899). Comptes rend. Acad. Sci. Paris. 128, 864.

HERTWIG, O. (1876). Morph. Jahrb. 1, 347.

HERTWIG, O. (1877). Morph. Jahrb. 3, pp. 1 and 271.

HERTWIG, O. (1878). Morph. Jahrb. 4, 156.

HERTWIG, O. (1890). Arch. f. mikro. Anat. 36, 1.

HERTWIG, O. (1893). Die Zelle und die Gewebe Vol. I. Jena.

HEUSER, E. (1884). Bot. Centralblatt. 17, 27.

HOFMEISTER, W. (1848). Bot. Zeit. 6, pp. 425, 649, 670.

HOFMEISTER, W. (1849). Die Entstehung des Embryos der Phanerogamen. Leipzig.

KROHN, A. (1852). Arch. f. Anat. u. Physiol., p. 312.

LOVEN, S. L. (1848). Ibid., p. 531.

MARK, E. L. (1881). Bull. Mus. Comp. Zool. Harvard. 6, 173.

MOHL, H. VON (1837). Allg. bot. Zeit. 1, 17.

MÜLLER, O. F. (1786). Animalcula infusoria fluviatilia et marina. Copenhagen.

NÄGELI, K. (1844). Zeit. f. wiss. Bot. 1, 34. Trans. (1846) A. Henfrey in Roy. Soc. Reports and Papers on Botany.

NAVASHIN, S. G. (1898). Bull. Acad. Sci. St. Petersb. 9, 377.

NAVASHIN, S. G. (1899). Bot. Centralblatt. 77, 62.

NEWPORT, G. (1852). Proc. Roy. Soc. London. 6, 171.

NEWPORT, G. (1854). Philos. Trans. Roy. Soc. London, p. 229.

OVERTON, E. (1893). Ann. Bot. 7, 139.

PRÉVOST, J. L., AND DUMAS, J. B. (1824). Ann. des Sci. nat. 2, 100.

PRINGSHEIM, N. (1855). Ann. Sci. Nat. (Bot.) 3, 363.

PRINGSHEIM, N. (1856). Ann. Sci. Nat. (Bot.) 5, 250.

RABL, C. (1885). Morph. Jahrb. 10, 214.

REICHERT, C. B. (1847). Arch. f. Anat. u. Physiol. Wiss. Med., p. 88.

REMAK, R. (1841). In Bericht über die Leistungen im. Gebiete der Physiologie, p. 17.

REMAK, R. (1852). Arch. f. Anat. u. Physiol. wiss. Med., p. 47.

REMAK, R. (1858). Ibid., p. 178.

ROBIN, C. (1862). J. physiol. de l'homme et des animaux. 5, 149.

SCHLEICHER, W. (1879). Arch. f. mikro. Anat. 16, 248.

SCHLEIDEN, M. J. (1837). Arch. f. Naturgeschichte. 3, 289.

SCHMITZ, C. J. F. (1879). Sitzber. Niederhein Gesell. Bonn, p. 345.

SCHNEIDER, A. (1873). Jahrb. Oberhess. Gesell. Natur.-Heilk. Giessen. 14, 69.

SELENKA, E. (1878). Sitzber. Phys. Med. Soc., Erlangen. 10, 1.

SPALLANZANI, L. (1786). Expériences pour servir a l'histoire de la génération des animaux et des plantes. Genève.

STRASBURGER, E. (1875). Zellbildung und Zelltheilung. 1st ed. Jena.

STRASBURGER, E. (1877). Jen. Zeit. 11, 435.

STRASBURGER, E. (1879). Ibid. 13, 93.

STRASBURGER, E. (1880). Zellbildung und Zelltheilung. 3rd ed. Jena.

STRASBURGER, E. (1884). Das botanische Practicum. . . . Jena.

STRASBURGER, E. (1888). Histologische Beiträge 1.

THURET, G. (1854). Ann. Sci. Nat. (Bot.). 2, 197.

VAUCHER, J.-P. (1803). Histoire des conferves d'eau douce. Genève.

WALDEYER, W. (1888). Arch. f. mikro. Anat. 32, 1.

WARNECK, N. (1850). Bull. Soc. impér. des Naturalistes de Moscou. 23, 90.

WHITE, M. J. D. (1945). Animal Cytology and Evolution. 1st ed. Cambridge.

WILSON, E. B. (1895). Science. 1, 69.

CHAPTER FOUR

Theories of Inheritance since Darwin, and of the Role of the Nucleus in Heredity

CHARLES DARWIN (1809–1882) wrote the *Origin of Species* without any reference to cellular structure in animals and plants; in the index of the book the word 'cell' is not to be found. When, however, in his *Animals and Plants under Domestication*, which appeared some nine years later, a theory of heredity was proposed, Darwin then first discussed to what extent current views on the formation and structure of tissues were relevant to the problems of inheritance. At that time in England, as we have seen, the doctrine of '*omnis cellula e cellula*' was by no means exclusively held. There were some who still believed that cells were produced in a formless blastema. 'As I have not especially attended to histology', Darwin said (1868, p. 370), 'it would be presumptuous in me to express an opinion on the two opposed doctrines.' Furthermore his theory of Pangenesis had been conceived long before the appearance of the *Cellular Pathology*.

Darwin supposed that the cells of the body gave off tiny units, which he called gemmules. The gemmules circulate through the body and may collect together in various places. They have the power of developing into new cells. In the gonads they form the germ-cells, each of which contains gemmules derived from all cells of the body at all periods of development, and even from past generations. The fertilised egg contains gemmules derived from both parents and also from remote ancestors, for the great majority of these bodies are passively transmitted through each generation. In development, each cell is formed from a single gemmule and differentiates into the type of cell from the region of the body from which this unit was in the first place derived. Only a small minority of the total number of gemmules in the embryo develop into cells, and thus there is a large reserve which may be called upon, either at later stages of development or in future generations. This hypothesis, advanced with characteristic modesty, seemed to Darwin to afford a possible explanation not only of development and regeneration but also of what were then accepted as the facts of heredity. If the limb of an Amphibian were removed, for instance, the necessary gemmules would collect in the regenerating area and develop into the appropriate cells and tissues and become arranged into the required pattern. The theory could also be made to explain the influence of the gonads on the remote parts and organs of the body:

> Thus when male animals first arrive at puberty, and subsequently during each recurrent season, certain cells or parts acquire an affinity for certain gemmules which become developed into the secondary masculine characters, but if the reproductive organs be destroyed . . . these affinities are not excited. . . . The curious case formerly given of a Hen which assumed the masculine characters, not of her own breed but of a remote progenitor, illustrates the connexion between latent sexual characters and ordinary reversion. [1868, vol. II., p. 399.]

Darwin's theory of Pangenesis thus covers a wide range of still accepted phenomena which are now understood to be due to various causes and to be produced by quite separate mechanisms. It was also, however, able to explain other matters such as graft hybridisation and the inheritance of acquired characters which since Darwin's time have grown more doubtful. In the minds of most biologists Pangenesis and Lamarckian inheritance became closely associated, and when in the 1880's the inheritance of acquired characteristics passed into disfavour, Darwin's theory suffered a still more complete extinction.

Darwin himself was very fond of Pangenesis. When the *Animals and Plants under Domestication* appeared, his main anxiety over the book was how his friends would react to the theory, but the enthusiasm of Lyell and Wallace only partially compensated for the coolness of Hooker and Huxley towards his 'beloved child'. To Lyell he wrote in 1867:

> I am particularly pleased that you have noticed Pangenesis. I do not know whether you ever had the feeling of having thought so much over a subject that you had lost all power of judging it. This is my case with Pangenesis (which is 26 or 27 years old)....

Attempts were made to put the theory to the test of experiment. Darwin's cousin Francis Galton (1822–1911) thought that the presence of gemmules in the blood could be tested by transfusing the blood of rabbits of different breeds and afterwards examining the effect on their hereditary characteristics:

> According to Darwin's theory, every element of the body throws off gemmules, each of which can reproduce itself, and a combination of these gemmules forms a sexual element. If so, I argued, the blood which conveys these gemmules to the places where they are developed, whether to repair an injured part or to the sexual organs, must be full of them. They would presumably live in the blood for a considerable time. Therefore if the blood of an animal . . . were largely replaced by that of another, some effect ought to be produced on its subsequent offspring. [Galton, 1908, p. 296.]

These experiments, although practised on an extensive scale, were without result. Darwin's comment on them was that the presence of gemmules in the bloodstream was not an essential element in the theory; he did not allow that 'Pangenesis had as yet received its death-blow, though from presenting so many vulnerable points its life is always in jeopardy' (F. Darwin, 1887, iii, p. 195). So diffuse a body presented no single vital spot and, as Beale said, 'will withstand every attack that may be made. Like many favoured hypotheses of these days, it can neither be proved to be true nor positively shown to be false, and it is open to anyone to ground his belief in the truth of this or other doctrines upon the fact that they have not been and cannot be disproved' (Beale, 1871).

The belief of G. J. Romanes (1848–1894) in Pangenesis was grounded in veneration of its author, to whom letters from the younger man invariably ended 'very sincerely and most respectfully yours'. From 1873, and for seven years onwards, Romanes devoted much of his time to the study of graft hybridisation in plants to see whether, as Darwin believed, the gemmules from a graft were able to cross over and affect the tissues of the host. 'Altogether', as he later said, 'I made thousands of experiments in graft hybridisation (comprising bines, bulbs of various kinds, buds and tubers), but with uniformly negative results' (Romanes, 1895, p. 143). Unfortunately this work was never given to the world, as the author did not consider that such conclusions merited publication.

In the next decade there were very great developments in the climate of opinion on heredity. No one brought forward any unassailable positive evidence for the inheritance of acquired characters, and the declining remnants of Lamarckian belief among many biologists were dispelled by the teaching of August Weismann, who took the lead in constructing fresh theories of heredity upon the foundations of the new knowledge of the microscopical behaviour of the cell in reproduction and growth.

August Weismann was born in 1834. He qualified in medicine and was engaged in practice until 1863, but then spent two months at Giessen working in Leuckart's laboratory, where he found research in microscopical biology so engrossing that his future career was then determined. He settled at Freiburg and obtained the Chair of Zoology at that university in 1866. His early studies were on the embryology and metamorphosis of insects, mainly among the Diptera. These researches were interrupted by failure of his eyesight. However, after some years he was able to resume his labours and then began to investigate the sexual cells of the Hydrazoon Coelenterates. This work suggested his later ideas on the separateness of the 'germ plasm'. His disability, however, led to his gradual withdrawal from practical microscopy, and so from 1870 his main concern became the general subject of variation. His early papers on this subject were translated into English in 1882, with an introduction by Charles Darwin himself. From that time Weismann was entirely occupied with theoretical interpretation of the growing volume of cytological observations. His views were set forth in numerous essays and lectures, which became widely known. During the later 1880's they exercised a very considerable influence in this country.

Whereas Darwin had been primarily concerned with suggesting a mechanism whereby organisms could change in character during the long course of many successive generations, Weismann's views may be said to begin with the necessity of accounting for the stability of reproduction in living creatures. How is this pattern of organic design formed afresh in each generation, and a virtual replica of the parent so produced? If Pangenesis or any similar theory was to be rejected, Weismann (1885) stated the possible alternatives:

> Now if it is impossible for the germ-cell to be as it were an extract of the whole body, and for all the cells of the organism to despatch small particles to the germ-cells, from which the latter derive their power of heredity; then there remain, as it seems to me, only two other possible . . . theories as to the origin of germ-cells. . . . Either the substance of the parent germ-cell is capable of undergoing a series of changes which, after the building up of a new individual, leads back again to identical germ-cells, or the germ-cells are not derived at all, as far as their essential and characteristic substance is concerned, from the body of the individual, but they are derived directly from the parent germ-cell.
>
> I believe that the latter view is the true one. . . . I propose to call it the theory of 'The Continuity of the Germ Plasm', for it is founded upon the idea that heredity is brought about by the transmission from one generation to another of a substance with a definite chemical and, above all, molecular constitution. [English trans. 1889, p. 167.]

Although there were examples in embryonic development where a very early distinction could be traced between the primitive germ-cells and the remainder such as Weismann himself had discovered in the Diptera, the idea of the continuity of the germ plasm does not necessitate this, for Weismann defines the 'germ plasm' as

> that part of a germ-cell of which the chemical and physical properties—including the molecular structure, enables the cell to become, under appropriate conditions, a new individual of the same species. [Ibid., p. 174.]

As soon as it had been shown by O. Hertwig (1876) and by Fol (1879) that in the final stages of fertilisation two nuclei joined together, it became probable that the germ plasm was to be found within the cell nucleus. When Strasburger in 1884 showed that in the flowering plants the male element appeared to be solely a nucleus without the middle-piece or tail of the animal spermatozoon, the conclusion seemed clear that the nucleus alone was concerned in heredity. Subsidiary arguments were drawn from the circumstances that the male and female pronuclei were equal in size, although they are derived from cells so widely disproportionate. Moreover, the influence of each parent in heredity is roughly the same.

So far, in 1885, was the argument developed. Seven years later, however, in *The Germ Plasm, a Theory of Heredity* (trans. 1893), Weismann could say much more. The 'hereditary substance' could then be identified with the chromosomes. In fertilisation, it was then known, an equal number of these bodies is contributed by each parent to form a new nucleus from which the development of a new individual proceeds. This identification was further strengthened by the consideration of the events which occur during the division of the nucleus, from which Weismann drew the inference that

> the complex mechanism for cell-division exists practically for the sole purpose of dividing the chromatin, and . . . thus the latter is without doubt the most important part of the nucleus. [*The Germ Plasm*, p. 26. English trans.]

To this the great majority of biologists would still agree, save for some special exceptions of direct cytoplasmic inheritance. Everything, however, that Weismann built upon it has long since been swept away. Since he misunderstood the nature of reduction division and regarded it as a throwing out of half the chromosomes, he was obliged to endow each chromosome with the whole complement of hereditary material. Otherwise, he argued, it would be a matter of chance whether the complete outfit of germ plasm for the species was included in the two half-sets of chromosomes which united at fertilisation. His ideas on development and the way in which the organisation of the germ plasm was gradually transferred to the embryo are entirely opposed to our present epigenetic outlook, and are as outdated as the peculiar terminology in which the postulated structure of the various units of the germ plasm was expressed. The 'Biophor' has long been extinct, and the 'Id' has acquired a new meaning in another science.

One of the results of the application of the experimental method to cytology was a demonstration that normal development in an embryo is dependent on the presence of a whole set of chromosomes. From this the conclusion was drawn that individual chromosomes possess qualities which differ from one to another. It was first shown by Oscar Hertwig and by Fol that if a large excess of spermatozoa is used in fertilising a batch of sea-urchin eggs, occasionally two will penetrate a single egg, and that the first cleavage figure in such an egg will be a highly abnormal mitosis, usually with four poles. Such an egg at once then divides into four blastomeres, and usually the later development of the resulting larva shows a varying degree of abnormality. Boveri (1903) demonstrated that in such monsters the nuclei varied considerably in size, often in such a way that there were four contiguous areas in which nuclei of one particular diameter were found. These areas corresponded to the original four blastomeres, each of which had received an irregular number of chromosomes. This initial error, Boveri concluded, was responsible for the abnormal development of the whole larva. From this time onwards cytology in relation to heredity became increasingly concerned with the single chromosome, rather than with the whole nucleus. Investigation necessarily concentrated on those species in which the chromosomes were particularly large, or in which some were distinguished either by shape or individual behaviour.

It was during the course of meiotic division that such observations were specially rewarding, and the way was opened for much detailed research of this kind by the suggestion of Henking (1891) that reduction division begins by an association of the chromosomes in pairs, a process which was termed 'synapsis' by Moore (1895). Rückert observed this behaviour in the elasmobranch egg in 1892. His interpretation was that homologous chromosomes derived in the first place from each parental nucleus were in conjugation. In this way an interchange of material between the partners could produce new hereditary combinations in the offspring. Thanks to differences in size between the homologous chromosomes in each sex, Montgomery (1901) was able to prove that in Hemiptera the conjugating chromosomes or 'bivalents' were in fact pairs derived from each parent. The word 'chromatid' is used for the single elements of the bivalent pair.

It is one of several coincidences in the history of the study of heredity and of its relations to cytology that this conclusion was reached within a year of the rediscovery of the work of Gregor Mendel (1822–1884) by three separate research workers. In 1900 de Vries (1848–1935), Correns, and Tschermak independently published accounts of their researches in hybridisation in which each author referred to the experiments of Mendel which since 1865 had laid forgotten. All three had grasped the concept of the unit character in inheritance. In England, Mendelism was introduced by William Bateson of Cambridge, who since 1900 had been devoted to the study of variation. In Mendel's original work, fundamental laws of heredity emerged from the manner in which a contrasted pair of unit characters, tallness and shortness in peas, were transmitted, an example now widely familiar. Both tall and short peas of appropriate strains, when self-fertilised, breed true. However, when the tall and the short are artificially crossed, the hybrid generation are all tall. This character is thus the dominant member of the pair. When the hybrids in their turn are bred together, of the resulting generation one-quarter each are true-breeding tall and short peas, while the remaining half are hybrid talls which resemble their parents. From these results Mendel drew the inference that in the formation of the germ-cells of the hybrids the characters 'tall' and 'short' were distributed equally among the gametes. Thus such a hybrid produces four kinds of germ-cells; of those which are male and female, equal proportions of each carry in some way the characters 'tall' or 'short'. If the chances of fertilisation of each kind by any other are uniform, then the observed ratio of the three types of progeny would be expected to emerge.

This concept of 'segregation', as it was called, of the characters in the development of the germ-cells reached cytologists just at a time when they were beginning to be familiar with the process as it is represented at the microscopical order of size. In 1866 no answering chord had then been struck when Mendel laid his results before Nägeli, who nevertheless was deeply concerned with the problems of evolution and heredity in those years which followed the appearance of the *Origin of Species* (Iltis, 1932, p. 182).

After 1901, however, the school of cytologists in the United States of America were at once able to see the special significance of hereditary segregation. In that year, Montgomery, in a detailed study of meiosis in the Hemiptera, had only been able to suggest that synaptic conjugation of the bivalents was a means of effecting 'a regeneration of the chromosomes'. Yet by 1902 W. A. Cannon, E. B. Wilson and his pupil W. S. Sutton (1876–1916) were all aware of Mendel's work, and to them came the opportunity of revealing the true significance of the intricate processes of the maturation of the germ-cells which the researches of the preceding twenty years had already disclosed. In so doing they achieved one of the decisive steps in the progress of the biological sciences (Cannon, 1903; Wilson, 1902; Sutton, 1902).

Sutton had chosen an unusually favourable species for his investigations. In the spermatocytes of *Brachystola magna*, the great 'lubber' grasshopper, there are eleven pairs of readily distinguishable chromosomes, together with one unpaired element, the significance of which we shall later discuss (*Plate XII, Fig. B*).

He wrote two papers in that year; in the first he describes the reduction division of the spermatocytes of *Brachystola,* and in the second enlarges on its significance. The second paper begins with five generalisations:

> (1) The chromosome group of the presynaptic germ-cells is made up of two equivalent chromosome series, and strong grounds exist for the conclusion that one of these is paternal and the other maternal.
> (2) The process of synapsis . . . consists in the union in pairs of the homologous members (i.e. those that correspond in size) of the two series.
> (3) The first post-synaptic or maturation mitosis is equational and hence results in no chromosomic differentiation.
> (4) The second post-synpatic division is a reducing division, resulting in the separation of the chromosomes which have conjugated in synapsis, and their relegation to different germ-cells.
> (5) The chromosomes retain a morphological individuality throughout the various cell divisions.

On this basis he was able to go further and study their individual behaviour:

> A more careful study was made of the whole division process, including the position of the chromosomes in the nucleus before division, the origin and formation of the spindle, the relative positions of the chromosomes and the diverging centrosomes, and the point of attachment of the spindle fibres to the chromosomes. The results gave no evidence in favour of parental purity of the gametic chromatin as a whole. On the contrary, many points were discovered which strongly indicate that the position of the bivalent chromosomes in the equatorial plate of the reducing division is purely a matter of chance, that is, that any

chromosome pair may lie with maternal or paternal chromatids indifferently toward either pole irrespective of the positions of other pairs—and hence that a large number of different combinations of maternal and paternal chromosomes are possible in the mature germ-product of an individual.

He then estimates the total numbers of possible combinations of various numbers of chromosome pairs and continues:

> The constant size differences observed in the chromosomes of *Brachystola* early led me to the suspicion, to which, however, a study of spermatogenesis alone could not confirm, that the individual chromosomes of the reduced series play different roles in development. The confirmation of this surmise appeared later in the results obtained by Boveri in a study of larvae actually lacking in certain chromosomes of the normal series, which seem to leave no alternative to the conclusion that the chromosomes differ qualitatively and as individuals represent different potentialities. Accepting this conclusion we should be able to find an exact correspondence between the behaviour in inheritance of any chromosome and that of the characters associated with it in the organism.

Furthermore, when germ-cells combine, each of which bears either of two alternative chromatids, *A* or *a*, the result will be similar to that which Mendel had obtained in crossing hybrid peas. Progeny of three types, denoted in this example by *AA*, *Aa*, and *aa*, will again be produced in the ratio 1:2:1. So Sutton came to a general conclusion, which he expressed in these words:

> Thus the phenomena of germ-cell divisions and of heredity are seen to have the same essential features, viz. purity of units (chromosomes, characters) and the independent transmission of the same; while, as a corollary, it follows in each case that each of the two antagonistic units (chromosomes, characters) is contained by exactly half the gametes produced.

In a further paragraph he anticipates developments in the study of Mendelian heredity which belong to the following decade:

> We have seen reason in the foregoing considerations to believe that there is a definite selection between chromosomes and allelomorphs [Bateson's term] or unit characters but we have not before inquired whether an entire chromosome or only a part of one is to be regarded as the basis of a single allelomorph. The answer must unquestioningly be in favour of the latter possibility, for otherwise the number of distinct characters possessed by an individual could not exceed the number of chromosomes in the germ-products; which is undoubtedly contrary to fact. We must therefore assume that some chromosomes at least are related to a number of different allelomorphs. If then the chromosomes permanently retain their individuality, it follows that all the allelomorphs represented by any one chromosome must be inherited together. On the other hand, it is not necessary to assume that all must be apparent in the organism, for here the question of dominance enters and it is not yet known that dominance is a function of an entire chromosome. It is conceivable that the chromosome may be divisible into smaller entities (somewhat as Weismann assumes) which represent the allelomorphs and may be dominant or recessive independently. In this way the same chromosome might at one time represent both dominant and recessive allelomorphs. [1903b, p. 240.]

It was particularly true of this branch of knowledge at that time that progress depended on the study of exceptions to the generalisations which were first formulated. As Bateson said some years later, in his inaugural lecture at Cambridge:

> Treasure your exceptions. . . . Keep them always uncovered and in sight. Exceptions are like the rough brickwork of a growing building which tells that there is more to come and shows where the next construction is to be. [B. Bateson, 1928, p. 324.]

To the rule that the chromosomes associated in pairs was the exception that an odd one among them did not, and from the study of this peculiarity came the understanding of the inheritance of sex. A decade before the rediscovery of Mendelism, Henking (1891), in the series of studies in which he discovered the conjugation of the chromosomes in meiosis, had noticed that in *Pyrrhocoris*, a Hemipteran, in the second spermatocyte division, a 'peculiar chromatin-element' lagged behind the separating chromosomes in anaphase, and then passed undivided to one pole of the spindle. Thus there are two groups of the resulting sperm cells; half are with and half are without this body, which Henking finally decided was a nucleolus. By the turn of the century a number of other observers had made similar observations on spermatogenesis in other insects. In 1902 McClung had identified the unpaired body in spermatogenesis as an accessory chromosome, and on the basis of its regular distribution to half the total number of spermatozoa had posed the question whether it was to be regarded as a determinant of sex. By 1905 E. B. Wilson had surveyed reduction division in Hemiptera sufficiently widely to draw this conclusion:

> The sexes of Hemiptera show constant and characteristic differences in the chromosome groups, which are of such a nature as to leave no doubt that a definite connection of some kind between the chromosomes and the determination of sex exists in these animals. [Wilson, 1905, p. 500.]

The difference between the two sets of gametes was found to take various forms. In some insects half the sperms had an X-chromosome, as it was called, and the other half were without it; in others, those without the X had a much smaller Y-chromosome. In each of these species, however, all the eggs have an X-chromosome. As the two kinds of sperm are formed in equal numbers, at fertilisation there are even chances of the X-bearing eggs uniting with spermatozoa either with or without the X. In the former event the zygote is female; in the latter, male. Thus sex itself was shown to be inherited in Mendelian fashion. Not always, however, is it the male which is 'digametic'—that is, which produces the two kinds of gametes; in Lepidoptera it was shown by Doncaster (1910, 1914) and by Seiler (1913, 1917) that the female is sex-determining. The same has been inferred on genetical grounds for birds, but all other known instances of sexual determination conform to the pattern which was first discovered.

A second exception to Mendel's generalisation was described by Bateson, Saunders and Punnett in 1905. This again led to fruitful study in later years. Mendel's principle of free assortment was that in the production of germ-cells all the possible combinations of allelomorphs should occur in equal numbers. Bateson, Saunders and Punnett, however, found that in some experiments on sweet-peas the factors for blue flowers and long pollen grains tended to be associated together to such an extent that the normal ratios of the offspring were much disturbed. This phenomenon they termed 'gametic coupling', but was later displaced by the term 'linkage'.

It was in the United States that intensive research on this feature of inheritance was begun with the fruit fly, *Drosophila melanogaster*, under Thomas Hunt Morgan (1866–1945), towards the end of the first decade of this century. He had by then published many papers in experimental embryology, as well as a few on genetical subjects, mainly in criticism of the Mendelian doctrines. At that time, however, he came to grasp something of the possibilities of the fruit fly as a subject for genetical research. This organism can readily be reared on a simple medium of over-ripe bananas, and a new generation is produced every fortnight. Laboratory studies on inbreeding of this species were begun by W. E. Castle in the first years of the century. In Morgan's hands the systematic study of the spontaneous varieties which appeared among the millions of flies bred under standard conditions soon provided ample material for the most intensive research on heredity that had ever been undertaken.

One of Morgan's first papers on *Drosophila* (1910) concerned the inheritance of the factor for 'white eye', which is recessive to the normal red eye of the wild type. The results of crossing a white-eyed male with a red-eyed female are different from those obtained when the cross is made the other way round. The explanation involves a special form of linkage between this factor for eye colour and the inheritance of sex. The factor and its opposite allelomorph are borne only on the X-chromosome. It later became clear that the Y-chromosome bears no factors whatever. This is an example of sex-linked inheritance of the same kind as occurs in the disease in human beings known as haemophilia, in which one of the factors in the clotting of the blood is absent. A woman may transmit the condition to her sons without showing it herself. Precisely the same explanation applies to both examples.

Many instances of linkage between hereditary factors in *Drosophila*, however, were found to be independent of sex. It was soon realised that there were four groups of factors which showed linkage with each other, but not with those of any other group. Only one of those groups showed sex linkage. As more and more factors were recognised in the study of mutations in the fruit fly, the clearer did the conclusion emerge that each linkage group could be associated with one of the four pairs of chromosomes in the somatic cells of *Drosophila*. One linkage group was sex-linked and one other was very small.

The next step was to show that not only did every factor belong to a particular chromosome but that it had its own place on that chromosome, along which all the other factors of the same linkage group could be placed in a linear order. The possibility of 'mapping' the chromosomes was realised as early as 1911. The relative positions on the chromosomes of a pair of factors, or 'genes' as they were called about that time, depended on the strength of the linkage between them. To quote from Morgan:

> By linkage is meant that certain factors that enter the cross from each parent remain together in subsequent generations, more often than they separate. For example, if in *Drosophila* yellow wings and white eyes have entered from one parent, and grey wings and red eyes from the other, the new (cross-over) combinations, yellow and red, grey and white, are less numerous than are the original linkage combinations, yellow and white, grey and red. The number of individuals (cross-overs) that result from this interchange, expressed as a percentage of the whole number of individuals, is called the cross-over value. Such a percentage indicates how often the linkage is broken. Thus, if crossing over between yellow and white is shown in 1 per cent

of the gametes, then 1 stands for the cross-over value of yellow and white. Conversely yellow and white have remained together (linked) in 99 per cent of the gametes. We speak of the linkage relations in such cases in terms of the cross-over values, here 1 per cent. [Morgan, 1919, p. 118.]

If the genes are linearly arranged along the chromosomes, and if crossing over is the result of the interchange of segments of the chromatids at the time when they lie side by side in synapsis, then, it was argued, the nearer two genes are to each other the less is the chance of their becoming separated by this interchange. The test of this hypothesis came when comparison was made of the cross-over values of more than two pairs of genes belonging to the same chromosome. It was found that these values often showed simple arithmetical relations. If the cross-over values between three pairs of genes were *a*, *b* and *c*, and if *a* plus or minus *b* was equal to *c* then it was inferred that these values were proportional to the distances separating these genes along the length of the chromosome. Where, however, genes are situated at relatively long distances apart, then their cross-over values tend to be less than the sum of those of the intermediate genes. This is because there may sometimes occur a second point of crossing-over between genes so far away from each other that they are near opposite ends of a chromosome. A 'double cross' of this kind will bring them back to their original relationships.

Thus by patient study of the results of large numbers of matings of individuals bearing particular mutations those belonging to each linkage group were placed in a linear order, and so a map of the genes in the chromosomes of *Drosophila melanogaster* was constructed. By 1915, in *The Mechanism of Mendelian Heredity*, Morgan and his principal collaborators gave a preliminary chart of the location of fifty genes for the four single chromosomes. Subsequently each new mutation which appeared was assigned to its place; the first chromosome alone is now known to contain some five hundred genes.

The achievements of Morgan and his school were all the greater because they studied both genetics and cytology side by side. For many years this joint attack on the problems of inheritance was confined to this school alone. In England, Bateson's studies on Mendelian genetics continued for twenty years without reference to the chromosome theory of heredity. Finally, in 1921 he visited Morgan; the immediate effect of this journey on Bateson's scientific views is shown in extracts from letters which he then wrote home:

> I can see no escape from capitulation on the main point. The chromosomes must be in some way connected with our transferable characters. About linkage and the great extensions I see little further than I did. . . . Cytology here is such a commonplace that every one is familiar with it. I wish it were so with us. . . . [B. Bateson, 1928, p. 143.]

In the 1920's there were two further developments of great importance in the study of mutations in *Drosophila*. It was found that rearrangements of genes were not confined to reciprocal interchanges between homologous segments of chromosomes, but that whole segments of chromosomes could occasionally break away from their normal positions and move elsewhere. This 'translocation' of a chromosome, or part of a chromosome, was itself found to have genetical effects. Bridges (1923) was one of the first to describe a mutation of this kind; it resulted from the migration of a section of the second chromosome to the third.

In the same year Little and Bagg showed that inherited abnormalities could be produced in mice by the action of X-rays. Two years later Muller and Dippel showed that the fundamental effect on chromosomes of appropriate dosages of radiation was to cause breakages in them. The broken ends might subsequently unite either in their previous arrangement, but more frequently in some abnormal position. A sundered fragment of a chromosome could readily be lost from a cell during nuclear division. The genetical effect of radiation is thus to increase the frequency at which mutations appear in an organism from the normal rate of 1 in 10^{-5} or 10^{-6} in each generation (Lea, 1946, p. 131) to figures greatly in excess of this value.

One of the results of these developments in the science of genetics is that indirect evidence is thus provided of the cytological behaviour of the chromosomes during the maturation divisions of the germ-cells. At some time during the normal stages of meiotic prophase, the bivalent chromosomes must twist round each other and exchange corresponding portions of themselves. From genetical evidence it was even possible to show when this interchange must occur, thanks to the discovery in 1917 by H. H. Plough that the cross-over value between pairs of genes is not an absolute constant but, amongst other variables, is a function of temperature. Plough found that cross-over values in *Drosophila* are at a minimum at about 22°C.; above and below this range the curve rises to maxima on either side at 13° and 31°. He was able to make use of this effect to determine when crossing-over must actually occur in the sequence of events in the maturation of the egg-cells. Females heterozygous for three characters in the second chromosome were reared at a high temperature and mated to the corresponding triply recessive males. The mated females were then kept at normal temperatures. As successive batches of eggs were laid they were separately reared. After hatching, the cross-over values for these particular genes were determined. At first these values were high, showing the effect of the high temperatures at which the flies had previously been maintained, during which the egg-cells had undergone maturation. However, in eggs laid ten days after the flies had been returned to normal temperatures the cross-over indices had dropped to their usual values. By studying the progress of oogenesis under these experimental conditions, Plough was able to show, both in this and in other experiments, that crossing-over must occur at the stage of conjugation of the chromosomes.

Several years before these experiments were made, however, direct cytological evidence was available that chromatids did twist round each other in meiotic prophase. In 1905 Janssens described the configurations of the bivalent pairs in the spermatogenesis of Amphibia (*Plate XII, Fig. D*) and in later papers suggested the possible genetical significance of the crossed segments of the chromosomes, the 'chiasmata' (Janssens, 1909). In subsequent years much attention was paid by cytologists to this aspect of meiosis. Observations of this kind demanded not only the utmost powers of resolution of the microscope but also choice of the most favourable material. In animals the most suitable groups are Amphibia and the Orthoptera; among intensive studies on spermatogenesis in the latter group, one of the earliest was that of D. H. Wenrich (1916) on the grasshopper *Phrynotettix magnus*. There were, however, difficulties which hindered the identification of the cytologists' chiasmata with the geneticists' cross-overs. In the first place cytologists found spermatocytes much easier to study than oocytes, and so the earlier cytological observations were all on chiasmata of male germ-cells. Unfortunately, as Morgan and his colleagues began to realise as early as 1914, there is no crossing over in the male of *Drosophila*, either in the sex chromosomes, of which the Y is genetically inert, or even in any of the autosomes. It was not until 1930 that a difference in behaviour of the bivalent pairs in meiosis in the two sexes of *Drosophila* was demonstrated; Huettner in 1930 showed that in the spermatogonial divisions the stages of diplotene and diakinesis were omitted, while Guyénot and Naville had shown in the previous year that oogenesis in *Drosophila* was entirely normal in these respects.

Apart from special considerations of this kind, there were, and still remain, difficulties in the interpretation of the chiasmata themselves, even in the largest and most readily studied examples. The chromosomal units at this stage each consist of four strands, formed by the splitting of the homologous pairs. Even today different opinions are still held on the precise manner in which they may twist together and which of the four units are to be found paired together at each stage. Nor are the maturation divisions of *Drosophila* by any means among the easiest to observe, and but for a most fortunate circumstance the correlation of the behaviour of the chromosomes with their genetical consequences would probably still have remained imperfect. *Drosophila*, however, belongs to the order of Dipterous insects, in which many of the tissues of the larvae have giant nuclei, in which, although mitotic division has ceased, the chromosomes remain visible and discrete. They are twenty or more times as long as the corresponding chromosomes of a normal metaphase plate and show a banded structure throughout their length. In the larva of *Chironomus* they were known as far back as the 1880's, and were described first by Balbiani (1881) and shortly afterwards by Carnoy (1884) (*Plate XII, Fig. F*). It was not until the 1930's, however, that the larval salivary chromosomes of *Drosophila melanogaster* were examined, so separate a branch of biology had the genetics of this species meanwhile become; however, Kostoff (1930) then suggested that the banding was an expression of the linear sequence of genes in each chromosome, and from 1933 onwards T. S. Painter, of the University of Texas, began an intensive study of each of the salivary chromosomes of *Drosophila* in turn. Very simple techniques were found to be adequate; the whole salivary gland was crushed between slide and coverslip in an aceto-carmine mixture which acted as both fixative and stain. The crushing burst the membrane of the giant nuclei and spread out the chromosomes in a manner suitable for detailed study. The fundamental conclusions which Painter at once drew from his first studies on them may be quoted from his first paper in this field:

(1) Each of the chromosomes has a definite and constant morphology and is made of segments, each of which has a characteristic pattern of chromatic lines or broader bands, which appear to run around the achromatic matrix. The same chromosomes, or characteristic parts thereof, may easily be recognised in different cells of an individual, or in different individuals of a species. If the position of one or more segments is shifted by some form of dislocation (translocation, inversion etc.) the exact morphological point (or points) of breakage can be determined and the segments identified in their new position. This discovery places in our hands, for the first time, a qualitative method of chromosome analysis and once the normal morphology of any given element is known, by studying chromosome rearrangements of known genetic character, we can give

morphological positions to gene loci and construct chromosome maps with far greater exactness than has been heretofore possible.

(2) In old larvae, homologous chromosomes undergo a process of somatic synapsis. This union is more than a simple apposition, for the elements pair up line for line in the most exact way, and form one apparent structure. If one of the homologues carries an inverted section we get typical inversion figures. If one of the homologues is deficient at some point, the two mates unite except at the point of deficiency where the normal element usually buckles. Thus we can readily determine exactly how much of the one chromosome is missing. . . . [Painter, 1933.]

This paper was illustrated by a drawing of the X-chromosome from a larval salivary gland alongside of which was placed the older type of chromosome map of the same element, showing the genes which it carries, placed in their relative linear order. It was thus possible to indicate the bands on the salivary chromosome which represent the actual genes themselves.

Thus it may be said that with Painter's achievement the chromosome theory of heredity reaches one of its primary objectives, the identification within the nucleus of the sites where the hereditary factors are located, and from whence they must exert their controlling power. The very success of this line of research, however, emphasises still further how little is known of the nature of this influence. In cellular organisms this great question confronts us with two groups of problems: first, what the gene actually is, and next how it exerts an effect within the single cell—in the first place, the fertilised egg-cell. Secondly, although an identical set of genes is distributed to every cell of the embryo, they nevertheless come to exercise separate influences on different types of cell and on the various organ systems of the embryo. It must be admitted that in the last seventy years the individual and apparently diverging developments in cytology and genetics on the one hand and in experimental embryology on the other have heightened the sense of mystery which surrounds the expression of heredity in the course of development.

Weismann's original theory of the germ plasm set out to solve both groups of problems at one and the same time. For him the germ plasm in the nucleus of the fertilised egg consisted of a complete blue-print of the whole future organism, to which he gave the name of the 'Id'; the appropriate parts and sections of this were distributed through the embryo as cleavage progressed. The idea was wholly preformationist, for, as E. S. Russell has said,

> To the developed organism there corresponds, point for point, the complex architecture of the Id. Development is brought about by the orderly disintegration of this complexity so that, as far as the Id is concerned, development is a process of simplification which proceeds *pari passu* with the visible increase in complexity of the organism itself. The two processes are exactly complementary. What Weismann has done is to give an inverted description of the process of development in terms of a purely hypothetical complexity. [Russell, 1930, p. 50.]

Among embryologists the whole trend of thought for more than a century has been away from such conceptions and towards the opposite theory of epigenesis, which maintains that the diversity of structure which is gradually manifest during embryonic life is in no sense present when development begins but is formed anew in each generation. So we are forced to regard the contents of the zygote nucleus as something of the nature of 'information', something which can ultimately be translated into the details of a pattern which has yet to be sketched out. How this is done we are only at the very beginnings of our attempts to understand; yet within the bounds of the single cell something has already been learnt of the way in which the genes may exert their effect. This development has emerged from studies on the chemical composition of the nucleus—a subject with its own history, of which some account must now be given.

The founder of this branch of chemistry is Fredrick Miescher

(1844–1895). His father had studied under Johannes Müller, and he himself was a pupil of Wilhelm His, at one time of Basle. Miescher's post-graduate work began in the late 1860's, under the leading physiological chemist of the time, Hoppe-Seyler. Miescher's choice of a problem, however, was entirely his own. By that time it was clear from its universal occurrence that the nucleus was a peculiarly important cell constituent. Before any chemical analyses of the nucleus were possible, however, it would first be necessary to separate in quantity nuclei from the remainder of the cell. This Miescher decided to attempt, and made what seems at first sight the bizarre choice of pus cells from discarded surgical bandages, as material on which to work. Pus was certainly plentiful enough in hospitals before the days of antiseptics, yet, as he said, it was 'nicht tadelfrei', not without objection. He took the precaution of rejecting those bandages which smelt too badly.

He found that it was possible to get the cells into suspension, and furthermore, by treatment with dilute hydrochloric acid, pepsin and ether, to separate the nuclei from everything else. They settled to the bottom of the flask, and under the microscope this sediment was seen to be composed of 'wholly pure nuclei, with smooth contour and uniform contents, each with a sharply distinct nucleolus, but slightly smaller in size than at first' (Miescher, 1871). At that time several workers isolated nuclei from tissues by similar methods (Brunton, 1870; Auerbach, 1874), but in later decades such procedures lay long neglected, until revived in the 1930's under the name of cell-fractionation (Claude, 1937).

From the nuclei of pus cells Miescher prepared a substance of remarkable properties, to which he gave the name of 'nuclein'. It was a stronger acid than any other known biological material and was further distinguished by a high content of phosphorus, an element then rarely found in organic substances of physiological origin. So remarkable did Miescher's results then appear that Hoppe-Seyler was reluctant to publish them until he had himself confirmed these conclusions, so the appearance of Miescher's paper was delayed by two years until 1871.

In that year, however, he returned to Basle, and found in the ripe testis of the Rhine winter salmon a further and no doubt more attractive source of nuclein. In the isolated heads of the sperm he found not only the acidic 'nuclein' or nucleic acid but a highly basic nitrogenous substance with which it was combined, to which he gave the name of 'protamin'. Nuclein was prepared from this source by dissolving the sperm heads in strong salt solution, from which fibres of the material were precipitated by dilution with water. It was necessary to keep these preparations as cool as possible, and so he worked in an unheated room in winter. He described his methods of work in a letter to a friend in these words:

> When nucleic acid is to be prepared, I go at five o'clock in the morning to the laboratory. . . . No solution can stand for more than five minutes, no precipitate more than one hour being placed under absolute alcohol. Often it goes on until late in the night. Only in this way do I finally get products of constant phosphorus proportion. [Greenstein, 1943.]

Miescher found that among the characteristic properties of nuclein were its resistance to peptic digestion and its solubility in alkali. On treatment with strong salt solutions it swelled and gelated. These tests Zacharias in 1881 applied to various kinds of cells and nuclei under the microscope. He found that pepsin digested away the cytoplasm of frog erythrocytes, leaving the isolated nucleus. The same occurred with Ciliata, such as *Vorticella* and *Paramoecium*, when the macronucleus remained undissolved after the experiment. These nuclei were soluble in alkali. Among plant tissues, he made the interesting observation that in pollen mother-cells in division the chromosomes resisted pepsin, but that the spindle was digested. Flemming had this work of Zacharias in mind when, in the following year, in his *Zellsubstanz, Kern und Zelltheilung* he gave a definition of 'chromatin', the substance which forms the 'framework' of the nucleus:

> . . . in view of its refractile nature, its reactions, and above all its affinity for dyes, is a substance which I have named chromatin.

> Possibly chromatin is identical with nuclein, but if not, it follows from Zacharias' work that one carries the other. The word chromatin may serve until its chemical nature is known, and meanwhile stands for that substance in the cell nucleus which is readily stained.

We now know that the term 'nucleic acid' stands for a class of substances. This falls into two divisions, both of which are found in the nucleus. One does not occur outside it. Furthermore, in the living state they are combined with a wide range of proteins which are usually basic in type, and of which the protamine of the ripe fish sperm is an extreme and unusual example.

Progress in the study of the chemistry of the nucleic acids proceeded rapidly after the discovery by Altmann (1852–1901) of a method for preparing them free from proteins. This was published in 1889. Within a few years three components of the nucleic acids were recognised. In addition to phosphoric acid there are organic bases of two types, the purines and pyrimidines, and thirdly a carbohydrate which is a pentose sugar. The main sources from which nucleic acids were prepared were yeast and the thymus gland of animals. There were certain differences between the products isolated from these two substances, which premature generalisation erected into fundamental distinctions between the nucleic acids of animals and plants. In 1914, however, this dichotomy was shown to be a false one. It was then shown by the biochemist R. Feulgen (1914) that the unstable carbohydrate of the thymus type of nucleic acid was not a hexose, as it had been hitherto regarded, but a pentose, which on gentle hydrolysis liberated an aldehyde, which could be detected by the usual reagent for this class of substance, the dye fuchsin decolourised by sulphurous acid. Ten years later this test was applied to sections of tissues under the microscope. Feulgen and his collaborator H. Rossenbeck (1924) were then greatly surprised to find that the nuclei of the wheat germ gave a strong reaction to this test, for this result showed that a nucleic acid of the thymus type could be found in plant cells.

We now know that the distinction between the types of nucleic acid is related to their distribution within the cells of both animals and plants. The 'thymus type' of nucleic acid is found only in the nucleus; its constituent carbohydrate is desoxyribose, and it is thus known as desoxyribo-nucleic acid, or, for short, DNA. The other type of nucleic acid is found alike in animals and plants, both in the cytoplasm and within the nucleus, where it is the main constituent of the nucleolus. Again from its carbohydrate, δ-ribose, is derived the name ribonucleic acid, or RNA. Furthermore, the two types of nucleic acid contain the same purine bases but differ with respect to one of their pyrimidines.

The full significance and potentiality of Feulgen's test, applied as a histochemical method, only slowly became understood. It was recognised as a specific reaction for DNA only when the extent of the preliminary hydrolysis was specified, for other constituents of tissues can liberate aldehydes after more prolonged disintegration. Cytologists did not begin to employ the Feulgen reaction until the late 1920's; in the succeeding decade the literature on its use became a prominent feature in the development of cellular biology at that time. It was applied to the chromosomes of the salivary glands of the Diptera much at the same time when Painter drew attention to their genetical importance (King and Beams, 1934; Bauer, 1935). The conclusion from these researches and others was irresistible: the bands of the salivary chromosomes where the genes are located contain desoxyribonucleic acid. This substance must therefore be the actual material basis of heredity. Miescher himself had speculated that the nuclein and protamine which he isolated from the sperm head might be concerned with inheritance, and had suggested that isomerism of these substances might provide the basis for hereditary variation. It was later doubted whether the nucleic acids in themselves were sufficiently complex to provide for all the variety of pattern which must be postulated in any physical basis of inheritance (Matthews, 1915; Schultz, 1941), though further knowledge of the structure of nucleic acids, incomplete though it is as yet, has robbed this argument of all force.

In the later 1930's a group of biologists in Stockholm under T. Caspersson began to apply another method of microscopical investigation to the study of the nucleic acids within the nucleus and the cell. These substances absorb ultra-violet light very strongly over a band of the spectrum in that region, at about 2600 Å; i.e. at a wave-length about half the average of visible light. Nucleic acids absorb at this wave-length in virtue of the ring structure of the constituent pyrimidines and purines which exists in two tautomeric forms, which are continuously changing from one to the other at a frequency which corresponds to the relevant wave-length in the ultra-violet.

Microscopy in the ultra-violet was first shown to be possible at the beginning of the century (Köhler, 1904), though the difficulties involved at that time were very formidable. Quartz alone had the necessary transmission at these wave-lengths. No glass whatever could be used for the optical system. Moreover, the image formed by the microscope could not directly be seen but was recorded by photography. The aim of ultra-violet photomicrography in the first place was to enhance resolution by decreasing the wave-length of the light employed. In the 1920's a group of English microscopists under J. E. Barnard developed the technique to a very high pitch of refinement and were able to study some of the larger animal viruses in this way, at magnifications of several thousand diameters.

Caspersson and his colleagues, however, began to use ultra-violet photomicrography as a means of studying the distribution of nucleic acids within the cell. The method had the great advantage that unstained material is used as the object, and thus the contrast in the resulting photomicrographs is directly due to the tissue components themselves and not to their affinity for a stain, the depth of which is largely dependent on the conditions of use. So the density of nucleic acids could readily be compared from one tissue to another.

Caspersson, however, went considerably beyond photomicrography at a single wave-length. With photoelectric cells and amplifiers of great sensitivity, used in conjunction with the ultra-violet microscope, he succeeded in measuring the absorption of the various constituents of a single cell at a range of wave-lengths, and thus to plot absorption curves for tiny particles of biological material, well beyond the range of the most refined microchemical methods. These curves gave information not only about the nucleic acids within the cell but also on the proteins, which have a rather less well-defined band of absorption, with a maximum at about 2800 Å. The use of the microscope to plot absorption spectra over a wide range of wave-lengths has since given a new impetus to the development of the reflecting microscope, the focus and operation of which is independent of wave-length.

The techniques used by Caspersson and his colleagues, like any others in scientific work, are not without their limitations. They cannot distinguish between the two types of nucleic acid, though this can readily be done not only by parallel use of the Feulgen reagent but also by experiments with specific enzymes which will remove either DNA or RNA from sectioned material. Again the possibilities of distinguishing proteins of different types by measurement of ultra-violet absorption on so small a scale are subject to grave limitations.

Subsequent reserve on the details of Caspersson's results, however, has not obscured his main conclusions drawn from the survey of many types of organisms, nor the importance which is attached to these generalisations. It was found that wherever cells or tissues are growing rapidly, there the density of nucleic acids within them is relatively high. This was found to be true equally of a colony of bacteria at the beginning of its phase of growth, of the cells of an actively malignant tumour, or in an early embryo (Caspersson, 1950). Evidence of this kind and of others thus points to the conclusion that nucleic acids have some biological function in the processes of synthesis within the cell; in a tissue where cells are frequently dividing, the division of the nucleus demands not only the formation of a new set of chromosomes which will contain the same quantity of DNA as did the parent nucleus at the same stage of the growth cycle, but also of enough proteins of whatever kind are necessary for both nucleus and cytoplasm in the daughter cells.

Not only do rapidly growing cells contain much nucleic acid

but also, it was found, the distribution of both kinds within the cell is then such as to suggest a general picture of the course of synthesis. Such cells have large and dense nucleoli. The nucleolus itself contains RNA, but is surrounded by granules of chromosomal DNA. Round the nuclear membrane is a layer of cytoplasm which contains much more RNA than is found farther out towards the boundary of the cell. Together with evidence from the distribution of proteins within the nucleus, this arrangement suggested that in the region of the nucleolus proteins are synthesised. These then migrate outwards through the nuclear membrane, at the surface of which the cytoplasmic RNA is built up. The idea that the nucleus is the centre of anabolic activity within the cell has often been advanced—first by Martin Barry in the 1840's; frequently the extrusion through the nuclear membrane of material of various kinds has been described in cytological literature. The work of the Stockholm School, however, has led to the recasting of this notion in the form that cellular synthesis primarily begins with the desoxyribonucleoproteins of the gene string, and that the genes themselves find their expression through the elaboration of specific types of protein material.

It is now nearly twenty years since these theses were first put forward. They offered to biologists in many different fields a new way of regarding cells of all kinds, one which it became urgent to submit to the test of practical experience. At that time it might not have been foreseen that, among the various branches of biology, studies on micro-organisms would be the first to yield decisive evidence in support of these views.

About 1940 work was in progress which linked the production of mutations in fungal spores with the properties of their constituent nucleic acids. A. Hollaender and C. W. Emmons (1941) developed methods by which suspensions of the spores of the pathogenic fungus *Trichophyton* could be irradiated with equal doses of ultra-violet radiation over a range of wavelengths. The irradiated samples were then grown in culture, and these cultures were scrutinised for mutations which differed from the normal colonies in respect to such features as the degree of pigmentation, and also in their rate and habit of growth. Thus the rate of mutation of this organism under the action of ultra-violet light could be plotted against the wave-length of radiation employed. It was found that the curve showed a marked increase in the region where the absorption of the nucleic acids was at a maximum.

Evidence of this kind for the view that nucleic acids are concerned with heredity and mutation, though valuable, is indirect. Research in another field of microbiology, however, has now yielded evidence of a direct change of one variety of bacterium into another under the influence of a chemically pure sample of DNA, yet which is endowed with the specific property of transforming one strain into another. The transformed variant is genetically stable and continues to propagate colonies of the new type.

The history of this development in microbiology begins in 1928 with the observations of a public health officer, F. Griffith. He was concerned with the various strains of *Pneumococcus* which he was able to isolate from cases of pneumonia. One of the distinctions between the several types is their manner of growth in culture; colonies of virulent strains have a glistening appearance and are known as 'smooth'. They are denoted by the letter 'S'. Non-virulent varieties do not show this feature and are called 'rough' or 'R' strains. It is now known that in the smooth variants a capsule of polysaccharide material surrounds each individual cell. Griffith found that an 'R' strain could regain both its virulence and its smooth form by innoculation into a mouse, together with a large dose of a virulent culture which had first been killed by heating.

Today, Griffith's observation is regarded as one of the same status and importance as the discovery that mutations can be induced by X-rays. Both belong to the same decade, yet their subsequent histories were at first very different. H. Ephrussi-Taylor (1951) has said in a recent review that

> while the experiments with X-rays orientated genetical research for the succeeding decade, Griffith's discovery exerted virtually

no influence on biological thought until nearly twenty years later, for the absence of sexual reproduction in bacteria was sufficient to discourage any geneticists from studying these induced transformations.

It was not until the newer conceptions had been reached of the role of the nucleic acids in biological synthesis and in heredity, and chemical techniques had reached their present levels of refinement, that the phenomenon of bacterial transformation was submitted to further study. In 1944, Avery, MacLeod and McCarty isolated the DNA from a smooth strain of *Pneumococcus* and were able to show that when added to a culture of the 'R' type the substance was able to effect the change into the virulent form. 'The active transforming material . . .', they were able to conclude, 'contains no demonstrable protein, unbound lipid, or serologically reactive polysaccharide, and consists principally, if not solely, of a highly polymerised viscous form of desoxyribonucleic acid.'

In recent years it has been shown by R. D. Hotchkiss (1951) that purified DNA from an appropriate source is able to effect a mutation of another kind in the same organism. This is concerned with resistance to the action of penicillin. If *Pneumococci* of either 'R' or 'S' strains are grown in the presence of this agent, only a very small residue of resistant mutants finally persists in the cultures. These remaining strains were gathered from mass cultures, and their DNA was separated and purified. Again it was found that this material was able to effect a permanent transformation of one type into another. The frequency with which a resistant mutant is produced from the non-resistant strain was thereby increased from the normal spontaneous value of up to five new bacterial cells in a million to a figure ten thousand times greater.

Our survey of this branch of biology began with the study of the details of structure within the nuclei of animals and plants. With the definitive link to the intensive study of heredity under T. H. Morgan, this twofold discipline of cytogenetics acquired a number of basic principles which were collectively known as the chromosome theory of inheritance. There then followed two further developments; first its biochemical aspects were explored, and secondly the biological basis of the whole complex was extended, until, as even this fragmentary account has shown, the genetics of micro-organisms is now one of its most important constituent elements.

By now the whole field extends across most of the conventional divisions of biology. Its centre of gravity has shifted along the scale of living beings in the direction of what at one time would have been regarded as its humbler divisions, among which their cells are proving more accessible to fundamental experiments than are those of more highly organised species. This trend has by now resulted in surprising developments. Most remarkably of all, classical genetical analysis of characters into groups of linked genes is now proceeding among the bacteriophages—bodies so small that an electron microscope is necessary to reveal their appearance. In such forms of life, without either sexual reproduction or, as yet, recognisable chromosomes, a number of examples of the crossing-over of mutant characters has been demonstrated. This development has clearly profound implications in cytogenetical theory, once a wider synthesis in this field is reached.

These developments, however destructive of the older tidiness of the structure of academic biology, we may regard as the supreme achievement of one of its once discrete divisions. Cytology has now lost itself among the others.

LITERATURE FOR CHAPTER FOUR

ALTMANN, R. (1889). Arch. Anat. Physiol. (Physiol. abt.), p. 524.

AUERBACH, L. (1874). Organologische Studien. Breslau.

AVERY, O. T., MACLEOD, C. M., AND MCCARTY, M. (1944). J. Exp. Med. 79, 137.

BALBIANI, E. G. (1881). Zool. Anz. 4, 637.

BATESON, B. (1928). Wm. Bateson. Naturalist. Cambridge.

BATESON, W., SAUNDERS, E. R., AND PUNNETT, R. C. (1905). Experimental Studies in the Physiology of Heredity. Reports to the Evolution Committee of the Royal Society.

BAUER, H. (1935). Naturwiss. 23, 475.

BAUER, H. (1936). Zeit. Zellf. und mikr. Anat. 23, 250.

BEALE, L. S. (1871). Nature. 4, 25.

BOVERI, T. (1903). Verh. Phys.-med. Ges. Wurzburg. 35, 67.

BRIDGES, C. B. (1923). Anat. Rec. 24, 426.

BRUNTON, T. L. (1870). J. Anat. Physiol. 2nd ser. 4, 91.

CANNON, W. A. (1903). Bull. Torrey. Bot. Club. 30.

CARNOY, J. B. (1884). La Biologie Cellulaire. Fasc I. Lierre.

CASPERSSON, T. O. (1950). Cell Growth and Cell Function. New York.

CLAUDE, A. (1937). Amer. J. Cancer. 30, 742.

CORRENS, C. (1900). Berichte d. Deutsche Bot. Gesell. 18, 158.

DARWIN, C. (1859). On The Origin of Species by Means of Natural Selection. London.

DARWIN, C. (1868). The Variation of Animals and Plants under Domestication. vol. 2. London.

DARWIN, F. (ed.) (1887). The Life and Letters of Charles Darwin. 3 vols. London.

DONCASTER, L. (1910). Proc. Roy. Soc. London (B). 82, 88.

DONCASTER, L. (1914). The Determination of Sex. Cambridge.

EPHRUSSI-TAYLOR, H. (1951). Cold Spring Harbor Symposia Quant. Biol. 16, 445.

FEULGEN, R. (1914). Zeit. für Physiol. Chemie. 92, 154.

FEULGEN, R., AND ROSSENBECK, H. (1924). Ibid. 135, 203.

FLEMMING, W. (1882). Zellsubstanz, Kern, und Zelltheilung. Leipzig.

FOL, H. (1879). Mém. Soc. Phys. et Nat. Genève. 26, 89.

GALTON, F. (1908). Memories of My Life. London.

GREENSTEIN, J. P. (1943). Sci. Monthly. 57, 523.

GRIFFITH, F. (1928). J. Hygiene. 27, 113.

GUYÉNOT, E., AND NAVILLE, A. (1929). La Cellule. 39, 27.

HENKING, H. (1891). Zeit. Wiss. Zool. 51, 685.

HERTWIG, O. [W.A.O.] (1876). Morph. Jahrb. I, 347.

HOLLAENDER, A., AND EMMONS, C. W. (1941). Cold Spring Harbor Symp. Quant. Biol. 9, 179.

HOTCHKISS, R. D. (1951). Ibid. 16, 457.

HUETTNER, A. (1930). Zeit. Zellforsch. 11, 615.

ILTIS, H. (1924). Trans. E. and C. Paul (1932). Life of Mendel. London.

JANSSENS, F. (1905). La Cellule. 22, 378.

JANSSENS, F. (1909). Ibid. 25, 389.

KING, R., AND BEAMS, H. W. (1934). J. Morphol. 56, 577.

KÖHLER, A. (1904). Zeit. wiss. Mikr. 21, pp. 129 and 275.

KOSTOFF, D. (1930). J. Hered. 21, 323.

LEA, D. (1946). Actions of Radiations on Living Cells. Cambridge.

LITTLE, C. C., AND BAGG, H. J. (1923). Amer. J. Roent. 10, 975.

MCCLUNG, C. E. (1902). Biol. Bull. 3, 43.

MATTHEWS, A. P. (1915). Physiol. Chem. 5, 186.

MENDEL, G. (1865). Verh. Nat. Verein. Brünn. 4, 3.

MIESCHER, F. (1871). Hoppe-Seyler med. chem. Untersuch. 4, 441.

MONTGOMERY, T. H. (1901). Trans. Amer. Phil. Soc. 20, 154.

MOORE, J. E. S. (1895). Ann. Bot. 9, 431.

MORGAN, T. H. (1910). Science. 32, 120.

MORGAN, T. H. (1919). The Physical Basis of Heredity. Philadelphia.

MORGAN, T. H., STURTEVANT, A. H., MULLER, H. J., AND BRIDGES, C. B. (1915). The Mechanism of Mendelian Heredity. New York.

MULLER, H. J., AND DIPPEL, A. L. (1926). Brit. J. Exp. Biol. 3, 85.

PAINTER, T. S. (1933). Science. 78, 585.

PLOUGH, H. H. (1917). J. Exp. Zool. 24, 147.

ROMANES, G. J. (1895). Darwin and after Darwin. Vol. II. Heredity and Utility. London.

RÜCKERT, J. (1892). Anat. Anz. 7, 107.

RUSSELL, E. S. (1930). The Interpretation of Development and Heredity. Oxford.

SCHULTZ, J. (1941). Cold Spring Harbor Symposia Quant. Biol. 9, 55.

SEILER, J. (1913). Zool. Anz. 41, 246.

SEILER, J. (1917). Zeit. Ind. Abst. Vererb. 18, 81.

STRASBURGER, E. (1884). Neue Untersuchungen über den Befruchtungsvorgange . . . Jena.

SUTTON, W. S. (1902). Biol. Bull. 4, 24.

SUTTON, W. S. (1903). Ibid. 4, 231.

TSCHERMAK, E. (1900). Berichte d. Deutsch. bot. Gesell. 18, 232.

VRIES, H. DE (1900). Ibid. 18, 83.

WEISMANN, A. (1882). Studies in the Theory of Descent, with a preface by C. Darwin. London.

WEISMANN, A. (1885). Essays upon Heredity . . . Authorised translation. Oxford, 1889.

WEISMANN, A. (1892). The Germ Plasm, a Theory of Heredity. Trans. W. N. Parker and H. Rönnfeldt. London, 1893.

WENRICH, O. H. (1916). Bull. Mus. Comp. Zool. Harvard. 60, 55.

WILSON, E. B. (1902). Science. 16, 991.

WILSON, E. B. (1905). Science. 22, 500.

ZACHARIAS, E. (1881). Bot. Zeit. 39, 169.

CHAPTER FIVE

History of the Study of Cytoplasm

WE COME now to the history of the study of the cytoplasm, of its texture and contents. Until recently it could be said of this branch of cytology that considerably less had been achieved with certainty than in the study of the nucleus and its components. Within the last few years, however, there have been two major developments. In the first place there has been a mutual permeation of microscopical and biochemical methods of investigation; secondly, the electron microscope has become a cytological instrument. The first of these developments is advancing the study of both the nucleus and the cytoplasm, though the benefit to the latter has been the greater. The biological revelations of the electron microscope, however, at present almost wholly relate to cytoplasmic texture and derivatives. It is not yet understood why the macromolecular structure of the nucleus and its components is still largely outside the scope of electron optics.

The first difficulty to be overcome in all microscopical investigations is the technical problem of the preparation of the material. When sufficiently adequate measures of this kind are available it is then necessary to decide how far their application has already distorted the arrangement and texture of the undisturbed living tissue with which the investigation begins. The first of all cytoplasmic investigations illustrates a difficulty of this kind, though one now largely overcome, which in the past has hampered research. Dujardin's original paper (1835) which called attention to the 'sarcode' of Protozoa and of a liver fluke is mainly concerned with its degeneration under lethal conditions.

These observations began with Protozoa which are found within the body-fluids of earthworms. Of these he chose two species, an Amoeba known to Dujardin as *Proteus tenax*, and a Ciliate, *Leucophra nodulata*. These organisms would survive only for a few minutes when suspended in water, after which there were extruded from them blebs of a clear gelatinous material, to which he gave the name of 'sarcode'. In the Ciliate the whole outer layer swelled up, and within this clear zone vacuoles appeared. These vacuoles Dujardin likened to those normally present within intact Ciliata, and which Ehrenberg regarded as 'stomachs'. Dujardin furthermore found that a liver fluke placed between glass plates in water after some hours gave off blebs of sarcode, which later became vacuolated.

In cellular organisms the first cells which were examined in anything approaching an intact state were neurones (G. R. Treviranus, 1816; Ehrenberg, 1833, 1836). In tracing the general history of the cell-theory (p. 40, above) mention was made of the important paper of G. Valentin (1836) on the microscopy of the nervous system (*Plate VIII, Fig. C*) and to his reference to the cytoplasm of the neurone as 'parenchyme'. This he described as 'a grey-reddish finely granular substance', containing 'small, dispersed, separate, round particles'.

A few years later, however, Robert Remak described a striated appearance in a nerve process. He reported that within the larger fibres of the ventral cord of the crayfish an axial bundle of hundreds of very fine fibrils could be seen, running in parallel sinuous courses. This appearance was evanescent, for soon after such fibres had been excised and prepared for examination the fibrillar appearance degenerated into a granular residue. In the year following, Remak (1844) published drawings both of the fibrillae in the nerve fibre and also of their appearance within the cell-body (*Plate VIII, Fig. D*). Here they followed a parallel course between nucleus and cell-body. To them the name of 'neurofibrillae' has since been given. There is still a great deal to be learnt about them.

Remak's observations passed without notice for over a decade. From the late 1850's onwards, however, there are several descriptions of a striated appearance within the cells of ciliated epithelia. Such a texture was described by N. Friedrich (1859) in the ependymal cells of the human brain, and separately by both P. Marchi and C. J. Eberth in 1866 in the ciliated cells of the gut and elsewhere in the fresh-water Lamellibranch *Anodonta*.

The scope of investigation of cells and tissues was very greatly extended by the introduction of methods of fixation and staining, though the possibilities of error were equally enlarged. The employment of silver salts for the impregnation of nerve cells was introduced by C. Fromman in 1864, and in the following year both he and J. Arnold described a more elaborate structure in the cytoplasm than had hitherto been observed. In the work of each of these authors there is illustrated a system of tubes or fibres leading from the interior of the nucleus through the cytoplasm and out into the nerve fibres. The particular form of this early tribute to the importance of the nucleus within the cell has not been confirmed by later studies on the structure of the neurone.

From this time onwards there are many accounts of a net-like texture of cytoplasm. Many, although not all, of them are based on fixed material, the appearance of which is often uncritically accepted as a representation of the living condition of the cell. Again, in much of this literature authors are interested in networks in both the nucleus and the cytoplasm, and describe continuous fibrils radiating from one into the other. Of this group, C. Heitzmann (1873) is among the first. E. Klein in 1878 minutely describes how networks can be seen in the cells of amphibian tissues after fixation in salts of chromic acid. His networks are not only continuous from nucleus to cytoplasm but extend on occasion from one cell to another. Flemming, in the *Zellsubstanz, Kern, und Zelltheilung* (1882) was much more cautious. His studies of cells, both resting and dividing, were also largely confined to those of Amphibia. Within the cytoplasm of living cartilage cells he clearly saw a system of discrete filaments; after fixation in osmic acid this appearance was retained, still with separate elements, not joined into a network. The texture of the living nucleus suggested to Flemming the term 'Kerngerüste', the nuclear framework. It was possible to preserve its form after fixation, provided that reagents containing chromic and picric acids were avoided.

In the 1880's studies on cytoplasm in many different organisms, both plant and animal, continued to describe reticular textures. These networks were interpreted in various ways. They were often regarded as a framework of solid material, enclosing a fluid component in the 'interstices within the intersections' of Dr Johnson's definition. To these two elements F. von Leydig (1885) gave the names of 'spongioplasm' and 'hyaloplasm'. J. Reinke and H. Rodewald (1881) attempted to separate two such constituents from the plasmodium of the slime-fungus *Aethalium* by means of pressure and centrifugal force, though without success. It is now known (Kamiya, 1942) that the constitution of the protoplasm of these organisms is greatly changed by treatment of this kind.

In 1878 Otto Bütschli suggested a further interpretation of the apparent reticular structure of protoplasm. This view was based on his observations on the structure of Protozoa:

> From the protoplasm of many Protozoa in which appear scattered vacuoles, there is a gradual transition to be found to completely alveolar, or what is the same thing, reticular protoplasm, where the alveoli are so densely crowded that their real protoplasmic walls take on a honeycombed arrangement, which in optical section appears reticular. [Bütschli, 1894, p. 2.]

This conception of protoplasm as an emulsion of two immiscible fluids was developed by Bütschli in a series of papers on the Protozoa, and finally in a book (1892, translated 1894), a large part of which is concerned with analogies in appearance between protoplasm and artificial emulsions. The fact that such mixtures can show something like amoeboid behaviour under appropriate conditions is, however, subsidiary to their apparent identity of texture with that of cytoplasm under the highest powers of the microscope.

At the end of the decade, however, powerful evidence was brought forward which indicated that the disperse appearance of protoplasm in histological preparations could be the result of the action of the fixatives employed in the first stage of treatment of the material. In 1899 appeared the *Fixierung, Farbung, und Bau des Protoplasmas* of Alfred Fischer, Professor of Botany at Leipzig, while in the same year Mr W. B. Hardy of Cambridge published a paper on the 'Structure of Cell Protoplasm'.

Hardy's work relates particularly to the reticular and alveolar theories of protoplasmic structure. The first section of the paper is on the nature of the changes produced in colloids by fixatives, and opens with these words:

> It is, I think, one of the most remarkable facts in the history of biological science that the urgency and priority of this question should have appealed to so few minds. Yet the urgency lies patent to the most superficial consideration. It is notorious that the various fixing reagents are coagulants of organic colloids, and that they produce precipitates which have a certain figure or structure. It can also readily be shown . . . that the figure varies, other things being equal, according to the reagent used. It is therefore cause for suspicion when one finds that particular structures which are indubitably present in preparations are only found in cells fixed with certain reagents, used either alone, or in particular formulae.

Hardy first showed by the simple application of pressure to a colloid how much its structure must be altered by the action of fixatives, and found that water could not be expressed from a gelatin gel until it had been treated with such reagents as formalin or mercuric chloride. The effect of fixation and staining on the microscopic appearance of diluted egg white was then studied. All fixatives with the exception of osmic vapour produced the texture 'of an open net with spherical masses at the nodal points'. The size of the meshes varied with the fixative and the concentration employed; mercuric chloride gave a particularly coarse result.

He then proceeded to fix a number of tissues with a range of reagents, and compared the appearance of each type of cell in the various series of experiments. Coarse networks again resulted from fixation with corrosive sublimate, while a much finer reticulum was seen in cells treated with osmic vapour (*Plate XII, Fig. H*). With this fixative the cells near the surface of a fragment of tissue showed a finer texture than those at deeper levels, to which the osmic vapour penetrated only slowly. Again the appearance of the reticulum in any specimen depended on the thickness of the section.

Hardy's conclusion from this series of experiments was as follows:

> There is no evidence that the structure discoverable in the cell-substance . . . after fixation has any counterpart in the cell while living. A large part of it is an artifact. . . . The framework which is visible in fixed cells contains within itself all the solids of the cell; it is produced by the action of the fixing reagent in converting the $\pm 10\%$ of solids in the living cell-substance into an insoluble state. The meshes of the framework are mere interstices occupied by alcohol, xylol, or balsam. . . .

Hardy was not the first to discuss the changes within the cell which are produced by fixatives. Flemming (1882, p. 51) had shown that in the cytoplasm of *Spirogyra* a fine network was produced by the action of osmic acid. Experiments similar to Hardy's with model substances furthermore had already been described; after Bütschli's book on 'Protoplasm and Microscopic Foams' had been written he found that films of egg-white or of gelatin when coagulated assumed an alveolar structure. A note on these observations was added to the book, without full discussion of their bearing on Bütschli's main thesis.

The demonstration that networks in cytoplasm could be produced by fixation did not, however, dispose of the reticular theory of protoplasmic structure, for there remained some apparent observations within the living cell. Such descriptions, however, are usually based on cells containing many vacuoles and other inclusions. For instance, a photomicrograph of a living Amphibian leucocyte made by S. Stricker in 1890, which was said to have convinced the histologist Schäfer of the existence of protoplasmic networks in such cells (Bütschli, 1894, p. 206) reveals on inspection a highly vacuolated cytoplasm, such as might be expected within any phagocytic cell. Yet by no means all of Bütschli's observations on alveolar structure in living cells can be dismissed in this way. However, concerning one of them a clear hint can be discerned from his criticisms of the observations of others. Bütschli correctly associates Flemming's inability to convince himself of the existence of cytoplasmic networks with the latter's use of the substage condenser at wide apertures (1894, p. 178). As the iris diaphragm is closed, Bütschli points out, a reticular texture in the image of a cell becomes more and more apparent. However, he wrongly regarded this as the true picture. Elsewhere (p. 86) he tells us that he used an apochromatic objective of N.A.1.4, but that

> The Abbé's condenser was used in some cases, but not in others, since I frequently remarked, as I thought, that finer structural relations came out more clearly without it. [1894, p. 86.]

The use of such an objective under these conditions is, as Baker (1942, p. 24) has remarked, 'like buying a Rolls-Royce and driving it uphill with the brakes on'.

The theories of protoplasmic texture which have so far been discussed do not, however, include all the different ways in which the structure of the cell substance has been regarded by cytologists. For an important group of observers protoplasm consisted of granules within an amorphous matrix. The larger categories of inclusions within the cytoplasm such as the plastids in plant cells and the secretory granules within glandular cells of animals have long been obvious. In the 1880's, however, several observers suggested that a finer granulation was an essential component of protoplasmic structure. The name 'microsomes' was suggested for these granules by Hanstein in 1882. In the same year Martin advanced the view that protoplasmic fibrillae might be composed of rows of such units and took up the suggestion of A. Béchamp (1868) that they might indeed be the ultimate elementary units of life—a hypothesis suggested by their resemblance to micrococci.

It is with Richard Altmann (1852–1901), however, that the development of this conception is mainly associated. In 1886 he described a method of staining cell granules with acid fuchsin, afterwards differentiated with picric acid—a procedure which is still of service. He described the granular texture seen in various cells after this treatment and suggested that these bodies played some part in the respiration of the cell. The theory that the granules are elementary units of life and homologous with bacteria was developed in a further work first published in 1890, with the title of *Die Elementarorganismen und ihre Beziehungen zu den Zellen*. The coloured illustrations in this work are superb and show how beautiful must have been Altmann's original preparations. Inspection of these plates reveals quite clearly that the various inclusions within cells which were stained by his technique were by no means all of one kind. No special significance can be assigned to their common affinity for fuchsin, nor is there any importance to be attached to the resemblance between a bacterial colony and granular cytoplasm when both are stained in this way.

Altmann's studies on cytoplasmic inclusions were deprived of much of their due influence at the time because of their association with his theories of elementary organisms. His work was, moreover, subjected to criticism of another kind. In 1894 Alfred Fischer began his series of researches on the effect of fixatives on colloidal solutions; their aim in the first place was to test the granular theory of protoplasmic structure. Fischer found, as did Hardy at the end of the decade, that fixatives produced granular or reticular textures in films of proteins. His most important discovery, however, was the effect of fixation on a mixture of two proteins in solution. When films of peptone and serum albumen were fixed by Altmann's mixture of osmic and chromic acids the two components became separated from each other; granules of peptone were then found within a matrix of coagulated albumen. In his later work (1899), moreover, where such experiments are described in full together with many others, the effect of staining such films was explored. Altmann's method was found to stain the peptone granules. Other stains, suitably differentiated, would colour them in other ways.

Among Altmann's granules, however, were to be found both wheat and chaff. In the closing years of the nineteenth century a series of papers were published by C. Benda of Berlin on the cytoplasmic inclusions of the developing spermatocyte (1898, 1899). Benda used a complex method of staining which involved haematoxylin, alizarin, and crystal violet. He was able to show that in the spermatozoa of the mouse the characteristic spiral filament of the middle-piece, which had first been described in the early 1880's (p. 22, above), could be stained by this procedure. Its formation was traced from single granules, which, in the later stages of development of the spermatozoon, aggregate into a continuous spiral (*Plate VI, Fig. C*). To cytoplasmic inclusions which could be stained by his methods Benda gave the name of 'mitochondria'—a name which recognises that they may be either granules or filaments.

In 1903 Benda showed that mitochondria were present in both eggs and sperm of the Amphibian *Triton*, and that after fertilisation they are handed on from cell to cell during cleavage. From observations of this kind Benda concluded that these inclusions are definite and permanent constituents of the cell. In the early years of this century the study of the mitochondria and their relations to the other contents of the cell attracted numerous workers; whereas in 1903 Benda found eighty-five relevant papers to survey in this field, by 1912 the bibliography of Duesberg's review contains upwards of six hundred items.

Among the most prominent contributors to this subject at that time was F. Meves. In 1908 he showed that mitochondria (or chondrioconts, as he alternatively termed them) can be seen throughout the tissues of vertebrate embryos. The threads in the cartilage cells of larval Amphibia, which Flemming had described in 1882, were mitochondria; so also were many of the cell inclusions which had been illustrated by Altmann in 1890, particularly those of a filamentous form. Altmann's method of staining with fuchsin and picric acid has since become one of the standard means of revealing mitochondria. In 1904 Meves demonstrated the presence of mitochondria in plant cells.

The acceptance of the mitochondria as a real constituent of protoplasm has been furthered by the relative ease with which they can be seen within the unfixed cell in suitable instances. We can now recognise that Flemming had observed them in untreated cartilage cells. Later, their identification was greatly aided by the introduction of methods of 'vital staining' at the turn of the century. Several dyes, such as methylene blue and neutral red, were found to be of such low toxicity that they could be applied to cells which were still alive. In 1900, however, L. Michaelis found that among these one dye, known as Janus green, could be used to colour one specific type of cell-inclusion. He found that in dilute solution this agent was taken up by the glandular cells of the pancreas and of the parotid of the mouse and stained numerous short threads therein, while leaving the secretory granules unstained. Janus green was used by Chambers (1914) to stain the mitochondria of the spermatocytes of a grasshopper, which, as in other insect spermatocytes, form a close investment to the mitotic spindle, and which at later stages of spermiogenesis form a coherent body which was first described by La Vallette St George in 1867 as the 'Nebenkörper'. It was within fibroblastic cells in tissue culture, however, that the most extensive observations on living mitochondria were made. The technique of tissue culture was introduced by Ross Harrison in 1907 in the first place as a means of studying the outgrowth of nerve fibres. He placed a fragment of the spinal cord of an amphibian larva upon a coverslip and surrounded it with clotted lymph, taking precautions to exclude micro-organisms. From the neurones of the 'explant' nerve processes grew outwards into the surrounding clot in much the same way as is seen in normal development. The 'outgrowth' from cultures of this kind forms a very thin layer of tissue which at the edge consists only of a single layer of cells, which are readily accessible to microscopic observation. Their movements and general behaviour can be directly followed.

This general technique was soon applied to the tissues of the higher vertebrates. Most mesodermal cells in the zone of outgrowth of such cultures take on a flattened fibroblastic form, in which long filamentous mitochondria are found within the cytoplasm. They were described by M. R. and W. H. Lewis in 1914. Ten years later, in a contribution to Cowdry's *General Cytology*, they illustrated the appearance of these bodies in numerous preparations stained with Janus green (*Plate XII, Fig. E*).

In the cytoplasm of cells of this kind, one other type of inclusion is to be found. This takes the form of small spherical droplets of fatty material, which together with the mitochondria are set in a matrix of clear cytoplasm which by all methods of light microscopy appears to be 'structureless'. It was found by Strangeways and Canti (1927) to be 'clear, transparent, and homogenous' when seen under the dark-field microscope. This sharp distinction between the inclusions of the cytoplasm and the clear background substance could not have been drawn at the time when W. B. Hardy was studying the effect of fixatives, and so at that epoch suspicion had fallen equally on all structures seen within fixed cells. Strangeways and Canti continued Hardy's line of enquiry and compared the appearance of the same cell in culture both before and after the application of various fixatives. It was found that only osmic acid preserved the cell in a fully life-like condition; furthermore their observations showed clearly that mitochondria are dissolved by fixation in fluids containing lipoid solvents.

Although the distribution and appearance of mitochondria in so many types of cell has been extensively studied over the last half-century, very little information of their function was available until recent years. The prominence of mitochondria in the blastomeres of early cleavage stages suggested to Benda that they must play some essential part either in fertilisation or heredity, or in cell-differentiation. It was at one time thought that mitochondria were formed only by division of pre-existing elements of the same kind, and in the zygote and in the subsequent embryo were all derived from the middle-piece of the spermatozoon. However, examples were later found where this part of the male element remained intact within one blastomere (Duesberg, 1919). The origin of mitochondria is still a largely unsolved question.

Altmann had originally suggested in 1886 that his granules were concerned with cell oxidation. Such a function was in later years ascribed to mitochondria, though not always on grounds which could now be accepted. It was argued, for instance, that organic solvents which remove mitochondria from a cell are also powerful narcotic agents and depress respiration. Of greater relevance, however, was the observation of G. W. Bartelmez and N. L. Hoerr (1933) of the prominence of both mitochondria at synaptic junctions round some large neurones of fishes, and of a copious blood-supply at these sites.

Clear evidence that respiratory enzymes are associated with mitochondria was provided in the first place by a method of investigating cells and tissues hitherto outside the general range of microscopical techniques. The separation of cell components in bulk from tissues was first practised in the early 1870's when Miescher prepared isolated nuclei (p. 99, above). During the 1930's several workers applied such methods to the cytoplasm. Tissues were mechanically disintegrated, and the resulting suspensions were resolved by differential centrifugation into fractions containing several distinct classes of cell components. In 1934 R. R. Bensley and N. L. Hoerr separated mitochondria in bulk from liver.

In the next decade it was shown that a number of important respiratory enzymes were associated with these isolated cell constituents, which on microscopical examination were shown to be unchanged in form and in affinity for Janus green (Hogeboom, Schneider, and Palade, 1948). This latter feature of mitochondria has since been shown by Lazarow and Cooperstein (1950) to be associated with their oxidative activity. They infer that the dye is reduced to a colourless derivative within the general cytoplasm, while at the mitochondrial surface it is kept in the oxidised form.

Attempts have also been made to explore the nature of the clear cytoplasmic ground substance by the methods of cell fractionation. In the differential centrifugation of cell suspensions the first component to settle out is the nuclei. They are followed by the mitochondria. The application of still greater centrifugal force then results in the sedimentation of fractions composed of granules of various sizes, which finally are of diameters of about a tenth of a micron. These have been termed 'microsomes' by Albert Claude (1943). They have been regarded as independently-reproducing elements within the cytoplasmic ground-substance; analysis shows them to contain a high proportion of phospholipids and ribonucleoproteins.

The use of the term 'microsome' for these granules is not the only feature in which contemporary exploration of the ultra-microscopic structure of cytoplasm recalls earlier researches on the composition of the cell. There have recently been both granular and reticular theories of protoplasmic ultra-structure. The evidence, however, for sub-microscopic networks in the cytoplasm is all indirect. Colloidal gels such as gelatin are known to possess a fine structure of this kind, which is responsible for mechanical properties such as their elasticity, and a viscosity which varies with the conditions of measurement. The consistency of protoplasm is extremely variable from cell to cell, but the more solid varieties of the living substance do show something of these gel-like properties. The analogy has at times been pushed to a surprising extent. The title of Prof. Frey-Wyssling's monograph of the 'Sub-microscopic Morphology of Protoplasm and its Derivatives' (1948) is printed against a background of an electron micrograph of a gel of vanadium pentoxide.

Little support for this conception of protoplasmic ultra-structure has been gained from the electron microscopy of the cell. Although the earlier photographs obtained by this means were capable of any possible interpretation, the later work in this field since the introduction of the ultra-microtome has begun to yield valid information concerning the ultimate texture of the cytoplasm. For instance there has recently been revealed a system of branched filaments, finer in calibre than are the mitochondria. These filaments are associated with granules which seem to contain much of the cytoplasmic nucleic acids. To this system the name 'ergastoplasm' has been applied, a term introduced by Garnier in 1900, with the meaning of a 'superior protoplasm', basophilic in reaction, which synthesises formed substances within the cell.

Within recent years the electron microscope has also begun to yield information of great importance concerning the internal structure of the larger mitochondria. Those of the mammalian liver and kidney were shown by Palade (1952) to possess an outer membrane which was infolded within the substance of the organelle to form a series of incomplete partitions or 'cristae'. In a number of other tissues mitochondria of a similar structure have recently been described, though at first sight so complex a make-up would seem in conflict with what is known of the mitochondria of the fibroblastic cell. These have long been recognised to be in constant slow movement, often accompanied by change of shape. It is not yet known, however, whether the larger mitochondria of the kidney tubule exhibit movements, nor whether those of fibroblasts are cristate in structure, though earlier electron micrographs of these cells have not suggested the presence of these features.

Not all the yet unsolved problems of the cell belong to the 'ultra-microscopic' order of size, however. There is one cell component which, when present, is well within the range of the ordinary microscope, the very existence of which has often been the subject of debate. This disputed structure is known as the 'Golgi-apparatus'. The name of the Italian neurologist Camillo Golgi (1844–1926) is associated therewith, because of his description in 1898 of a network of material impregnated with silver within Purkinje cells of the cerebellum of an owl. Golgi is one of the founders of the microscopical study of the nervous system; the methods of silver impregnation which he devised for this purpose were in use from the 1870's.

Golgi used the term 'internal reticular apparatus' for the network which he discovered in 1898, and expressed himself with great caution on its possible significance. In 1902 Kopsch demonstrated similar networks in spinal ganglionic neurones of various vertebrates by prolonged treatment with 2 per cent osmic acid. In subsequent years, networks of either silver or osmium were described in a variety of cells of vertebrate animals after suitable treatment. Sinigaglia in 1910 succeeded in producing a reticulum of this kind within the erythrocytes of the frog.

One of the many difficulties in the study of this 'apparatus' has been its confusion with other features within the cytoplasm. In 1900 E. Holmgren described a system of canals within the neurones of the spinal ganglia of a number of vertebrates. He suggested that these spaces represented the Golgi apparatus before impregnation, and that by the deposition of silver upon their walls a black network was thereby produced. This conclusion was accepted by the great neurologist Ramon y Cajal, who in 1907 began to speak of 'Golgi-Holmgren canals'. This identity, however, has not been sustained by later research. W. G. Penfield (1921) was able to show that when spinal ganglion cells were treated so as to reveal the Golgi network by Cajal's silver method subsequent bleaching and staining with iron haematoxylin revealed an entirely different system of canalicular spaces.

In the 1920's, however, a number of French cytologists claimed that vacuoles within secretory cells were the real basis of the Golgi body. This school maintained that cytoplasmic inclusions were generally of two kinds, either mitochondria or vacuoles which can be revealed by means of the vital dye, neutral red (Accoyer, 1924). Parat and his colleagues, in a series of subsequent papers, used the terms 'chondriome' and 'vacuome' for these two components of the cell. They maintained that lipoidal material of the chondriome surrounds the neutral red vacuoles and is responsible for the precipitation and reduction of silver or osmium around them (Parat, 1928). By further accretion this deposit may be joined into a continuous network. Thus, in their view, is the Golgi reticulum formed during the process of impregnation. The main criticism which has been levelled against this conclusion, however, is that not always can it be shown that vacuoles stainable with neutral red occupy the position of the Golgi body (Avel, 1925). Generally speaking, it is in the cells of vertebrates that networks have been demonstrated, while among the Invertebrata the Golgi material usually takes the form of separate elements, the 'dictyosomes'. This conclusion was reached as early as 1912 by R. Weigl, who thus showed that a Golgi body could not positively be identified by shape alone.

A symposium on the Golgi apparatus was recently held by the Royal Microscopical Society in 1954. The participants were divided between believers in the reality of this structure and those who 'admit no such thing'. It is not easy to come to any definite and final conclusion on this question by studying the papers which were then read. It does seem, however, that the burden of proof on certain fundamental questions still rests with the former group. Do the Golgi techniques in cytology necessarily reveal only structures which are homologous, and how constant in form are these bodies within the unfixed cell? To accept detailed descriptions of granules, networks, and vacuoles as wholly relevant thereto seems to demand an improbably static view of living protoplasm.

Among recent studies with the electron microscope there have been described vacuolar inclusions within the cell which from their position and their association with lipoidal material have been identified with the individual elements of the Golgi system. The study of such electron micrographs, however, suggests that the conception of separate categories of cell constituents, mitochondria, Golgi bodies and protoplasmic ground substance must soon be abandoned. There is great variety in form of the mitochondria, particularly where cells are active in metabolism; some of these variants are closely associated with the Golgi bodies. Moreover, in all the various inclusions of the cytoplasm, what are termed 'double membranes' enter into their composition; at boundaries within the cell, whether between the nucleus and cytoplasm, or at the cell wall itself, double lines can be traced within the most detailed electron micrographs which have so far been obtained.

It is probable that in a few years the cytology of the cytoplasm, and indeed much else in microscopical anatomy, will undergo a very extensive revision. Already, in discussing comparatively recent work on cytoplasmic structure it is necessary to bear closely in mind the period to which observations belong, and the methods by which they were made. This review of the history of cytology may thus belong to the dawn of a new period in the development of the subject, before which much of the existing content of the subject may soon be only of historical significance.

LITERATURE FOR CHAPTER FIVE

ACCOYER, H. (1924). Compt. Rend. Soc. Biol. 91, 665.
ALTMANN, R. (1886). Studien über die Zelle. Leipzig.
ALTMANN, R. (1890). Die Elementarorganismen und ihre Beziehung zu den Zellen. Leipzig.
ARNOLD, J. (1865). Arch. f. path. Anat. 32, 1.
AVEL, M. (1925). Compt. Rend. Soc. Biol. 93, pp. 26, 161.
BAKER, J. (1942). Some Aspects of Cytological Technique. In Cytology and Cell Physiology. Ed. G. Bourne. Oxford.
BARTELMEZ, G. W., AND HOERR, N. L. (1933). J. Comp. Neurol. 57, 401.
BÉCHAMP, A. (1868). Compt. Rend. Acad. Sci. Paris. 66, 366.
BENDA, C. (1898). Anat. Anz. 14, 264.
BENDA, C. (1899). Arch. Anat. Physiol. (Physiol. Abt.), p. 376.
BENDA, C. (1903). Ergeb. Anat. Entw. 12, 743.
BENSLEY, R. R., AND HOERR, N. L. (1934). Anat. Rec. 60, 449.
BUTSCHLI, O. (1892). Untersuchungen über mikroskopische Schaüme und das Protoplasma. Trans. E. A. Minchin, 1894. London.
CHAMBERS, R. (1914). Science. 40, 824.
CLAUDE, A. (1943). Science. 97, 451.
DUESBERG, J. (1912). Ergeb. Anat. Entw. 20, 567.

DUESBERG, J. (1919). Biol. Bull. 36, 71.

DUJARDIN, F. (1835). Ann. Sci. nat. (Zool.) 4, 343.

DUJARDIN, F. (1838). Ann. Françaises et Etrangères d'Anat. et de Physiol. 2, 379.

DUJARDIN, F. (1839). Ibid. 3, 65.

EBERTH, C. J. (1866). Arch. f. path. Anat. 35, 477.

EHRENBERG, C. G. (1833). Ann. Phys. Chem. (Poggendorff). 28, 449.

EHRENBERG, C. G. (1836). Abh. Akad. Wiss. Berlin, 1834, p. 605.

FISCHER, A. (1894). Anat. Anz. 9, 678.

FISCHER, A. (1899). Fixierung, Färbung und Ban des Protoplasmas. Jena.

FLEMMING, W. (1882). Zellsubstanz, Kern, und Zelltheilung. Leipzig.

FREY-WYSSLING, A. (1948). Sub-microscopic Morphology of Protoplasm and its Derivatives. Trans. J. J. Hermans and M. Hollander. Amsterdam.

FRIEDRICH, N. (1859). Arch. f. path. Anat. 15, 535.

FROMMAN, C. (1864). Arch. f. path. Anat. 31, pp. 129 and 151.

FROMMAN, C. (1865). Ibid. 32, 231.

GARNIER, C. (1900). J. Physiol. Path. gen. 2, 539.

GOLGI, C. (1898). Arch. Ital. de Biol. 30, 278.

HANSTEIN, J. VON (1882). Botanische Abteilungen. Bd 4, Heft 2. Bonn.

HARDY, W. B. (1899). Journ. Physiol. 24, 158.

HARRISON, R. (1907). Anat. Rec. 1.

HEITZMANN, C. (1873). Sitzber. d. K. Akad. d. Wiss. Wien. Abt. 3, 67, p. 100.

HOGEBOOM, G. H., SCHNEIDER, W. C., AND PALLADE, G. E. (1948). J. biol. Chem. 172, 619.

HOLMGREN, E. (1902). Anat. Anz. 18, 290.

KAMIYA, N. (1942). in The Structure of Cytoplasm, Ed. by W. Seifritz, Iowa.

KLEIN, E. (1878). Quart. J. Micro. Sci. 18, 315.

KOPSCH, F. (1902). Sitzber. K. Preuss. Akad. Wiss. Berlin. 40 (2), 929.

LA VALETTE ST GEORGE, A. DE (1867). Arch. f. mikr. Anat. 3, 263.

LAZAROW, A., AND COOPERSTEIN, S. J. (1950). Biol. Bull. Wood's Hole. 99, 322.

LEYDIG, F. VON (1885). Zelle und Gewebe. Bonn.

LEWIS, M. R., AND LEWIS, W. H. (1914). Science. N.S. 39. 330.

LEWIS, W. H. and LEWIS, M. R., in General Cytology, Ed. by E. V. Cowdry, Chicago, 1924.

MARCHI, P. (1866). Arch. f. mikr. Anat. 2, 467.

MARTIN, H. (1882). Arch. d. Physiol. norm. et pathol. 10, 465.

MEVES, F. (1904). Ber. Deutsch. Bot. Gesell. 22, 284.

MEVES, F. (1908). Arch. Mikr. Anat. 72, 816.

MICHAELIS, L. (1900). Arch. f. mikr. Anat. 55, 558.

PALADE, G. E. (1952). Anat. Rec. 114, 427.

PARAT, M. (1928). Arch. Anat. Micr. 24, 73.

PENFIELD, W. G. (1921). Anat. Rec. 22, 57.

RAMON Y CAJAL, S. (1907). Trab. del Lab. de Invest. Biol. T.5.

REINKE, J., AND RODEWALD, H. (1881). Untersuchungen aus dem. botan. Inst. der Univ. Göttingen. Hefte 2.

REMAK, R. (1843). Arch. f. Anat. Physiol. wiss. Med., p. 197.

REMAK, R. (1844). Ibid., p. 463.

SINIGAGLIA, G. (1910). Arch. Ital. de Biol. 53, 392.

STRANGEWAYS, T. S. P., AND CANTI, R. G. (1927). Quart. J. Micro. Sci. 71, 1.

STRICKER, S. (1890). Arbeit. aus Inst. f. allgem. u. exp. Path., Wien.

TREVIRANUS, G. R. (1816). In Treviranus, G. R. and L. C., Vermischte Schriften. 1, 117.

VALENTIN, G. (1836). Nov. Acta. phys.-med. Acad. Leop. 18, 51.

WEIGL, R. (1912). Bull. Acad. Sci. Cracovie, p. 417.

CHAPTER SIX

Cellular Theory in General Biology

In the preceding pages the history of the study of the cell has been reviewed under the separate headings of nucleus and cytoplasm. Till now we have discussed the cell-theory only in so far as it concerns historical views on the formation and reproduction of cellular units. Here we shall consider some of the wider implications of the doctrine and trace their impact on adjacent fields of biological enquiry. We must, however, remember that the present rigidity of the division of biology into separate sciences belongs to later phases of development of these studies, previous to which the distinction between various categories of investigators, such as embryologist, protozoologist, and the like, is largely irrelevant. The founders of microscopical biology, however handicapped by technical limitations, were at any rate free to examine the whole realm of living nature which lay before them.

In the second paragraph of Schleiden's *Beiträge zur Phytogenesis* he stated an important principle which has on many occasions been the subject of discussion:

> Each cell leads a double life: an independent one, pertaining to its own development alone; and another incidental, in so far as it has become an integral part of a plant. It is, however, easy to perceive that the vital process of the individual cells must form the very first, absolutely indispensable fundamental basis, both as regards vegetable physiology and comparative physiology in general; and therefore, in the very first instance, this question especially presents itself: *how does this peculiar little organism, the cell, originate?* [1847, p. 231.]

Here we see in the first place that the cell-theory is more than a statement that organisms are made up of cells; this proposition by itself had been an observation rather than a hypothesis well before Schleiden's day. From this, however, Schleiden drew two theoretical deductions, which may be re-stated in these terms:

(1) A cell is an individual unit, an 'elementary organism', as Max Verworn was later to term it.

(2) The cell is the seat of all vital processes, the investigation of which must necessarily, at some stage, be brought to the cellular level.

The idea of the cell as an individual was taken up with peculiar enthusiasm by Rudolf Virchow. Recent writers (Hirschfeld, 1929; Temkin, 1949; Ackerknecht, 1953) have emphasised how close was the parallel between Virchow's advanced liberal political views and his conception of the relationships between the cells and the body. As Ackerknecht says (p. 45), 'Cellular pathology showed the body to be a free state of equal individuals, a federation of cells, a democratic cell-state. It showed it as a social unit composed of equals. . . .'[1]

Whatever may be the ultimate value of such a parallel, its adoption concentrates attention upon the cell as a discrete unit, and away from the individuality of the whole organism. It was this aspect of Virchow's doctrines which was criticised in a review by Reichert in 1855. Virchow, he said, had constructed the whole animal from cells, as if they were atoms composing an inorganic body.

In the previous year Thomas Henry Huxley (1853) assumed an attitude towards this question completely opposed to the conception of the individuality of the cell:

> [. . . the cells] are not instruments, but indications . . . they are no more the producers of the vital phenomena than the shells

[1] The analogy between a cellular organism and society has been of service to other authors. It is to be found in Herbert Spencer, and in more recent years has been developed to an astonishing extent by Morley Roberts (1938), for whom an army is national ectoderm and the policeman the social phagocyte.

scattered in orderly lines along the sea beach are the instruments by which the gravitative force of the moon acts upon the ocean.

Although founded upon a completely false view of the metabolic function of intercellular material and heavily impregnated with Huxleyian rhetoric, this review later received high praise from both Michael Foster and Ray Lankaster (L. Huxley, 1900 (I), p. 140), by which time the tide of biological opinion, particularly in England and America, had for some years been running strongly against the idea of the individuality of the cell.

This movement owed much of its impetus to Adam Sedgwick of Cambridge, pupil and successor of F. M. Balfour, and a vigorous critic of the accepted general biological principles of his day. His analysis (1894) of von Baer's 'Law' is still the basis of our present views on this matter (de Beer, 1940). Early in the 1880's, after his teacher's untimely death, Sedgwick continued the work which Balfour had begun on the development of the primitive Arthropod *Peripatus*. By 1885 he had formed the opinion that the process of segmentation in the ovum of this creature did not result in the formation of discrete cellular elements. Sedgwick stated his conclusions in these terms:

1. The embryo at the gastrula stage and in all the earlier stages of development, is a syncytium.
2. No part of the nucleus of the unsegmented ovum enters the clear endoderm masses.
3. The solid gastrula consists of a multi-nucleate, much vacuolated mass of protoplasm.

Later, Sedgwick (1895) launched a general attack on the cellular theory of development, which, he said, 'blinds men's eyes to the most patent facts, and obstructs the way of real progress in the knowledge of structure'. By then Sedgwick brought forward evidence from elasmobranch embryos, mainly with respect to their mesenchyme and neural crest. About the former component he said that

This tissue is always described as consisting of branched cells lying between the ectoderm and the endoderm. The cells are spoken of as being separate from one another, and from the adjacent ectoderm and endoderm, excepting at points where they are supposed to arise from one of the primary layers. And not only are they described as being separate cells, but they are actually drawn . . . as separate from each other. This is, perhaps, the best instance that can be given of the bondage in which the cellular theory holds its votaries. For what are the facts? The separate cells have no existence at all! In their place we find, on looking into the matter, a reticulum of a pale non-staining substance holding nuclei at its nodes. It is these nodes, with their nuclei, which are drawn by authors as the separate branched cells of the mesenchyme, and they are constrained by this theory, with which their minds are saturated, not only to see things which do not exist, but actually to figure them.[1]

The rhetoric of the Cambridge zoologist was answered with studied restraint by an Oxford botanist, Mr G. C. Bourne. This debate served not only to exemplify the rivalry of the two ancient universities but also to illustrate the several points of view of animal and vegetable biologists. By that time there had been marked changes in opinion on the individuality of the cells of plants. In 1866 Ernst Haeckel had claimed that a fundamental difference between animals and plants was that only in the latter were the cells to be regarded as separate units. In the 1870's and 80's, however, numerous examples were described in plant tissues of protoplasmic connections between adjacent cells, now known by Strasburger's term 'plasmodesmata' (Strasburger, 1901). The literature on this topic has been admirably reviewed by Baker (1952) in the third of his series of papers on cellular theory. However, even after the recognition that protoplasmic connections between plant cells were probably of general occurrence botanists were unwilling to abandon the general concepts of cellular organisation. W. Gardiner (1884), one of the most prominent workers in this particular field, stated that 'the presence of minute perforations

[1] Sedgwick's views on intercellular connections in the embryo were cited by Herbert Spencer (1893) as evidence against Weismann's doctrine of the separateness of the germ plasm, and hence against the rejection of Lamarckian inheritance in neo-Darwinism.

of the cell-wall need not lead to any modification of all general ideas as to the mechanism of the cell'. Bourne himself did not wish to be freed of one dogma at the cost of subjection to an equally rigid but opposite view of Sedgwick,

> who would fetter us once more with a new doctrine, viz. there is no cell and thus all organisation is a specialisation of tracts and vacuoles in a continuous mass of vacuolated protoplasm.

A riposte by Sedgwick in the same volume of the *Quarterly Journal* added little further to the progress of the debate.

It is incumbent at this point to re-examine the evidence which Sedgwick urged against the cell-theory, and to discuss what further light upon these questions of histological structure has been revealed by the past sixty years. The embryology of *Peripatus* and related Onychophora has been re-examined by Manton (1949). The early development of these forms shows a peculiar feature in that before cleavage the egg suddenly swells in volume, and the incoming water separates the cytoplasm into a number of separate portions, only one of which contains the zygote nucleus. Solely from this is the embryo ultimately formed, and the remaining cytoplasmic spheres later degenerate; Sedgwick was wrong in regarding them as the source of the endoderm. His mistake was easily made, but it is unfortunate that he chose to place such emphasis on so untypical a form of embryonic cleavage.

On the general question of cell-boundaries we are now in a position to state that purely negative evidence of the inability to distinguish them in routine histological preparations cannot by itself be taken as proof of their absence. At the present time, when our understanding of the nature of animal tissues is being so rapidly enlarged by the use of the electron microscope, we can already form some new and general conclusions about cellular membranes. It is certainly true that even under the electron microscope the intercellular boundary is often inconspicuous. It is, in fact, no more substantial than is that of the nucleus or even the system of 'double membranes' which are so prominent a component within the cytoplasm. Yet in such photographs the distinction between a cellular tissue and a syncytium—as, for instance, between the cytotrophoblast of the human placental villi and the adjacent syncytiotrophoblast—is unmistakeable (Boyd and Hughes, 1954; Wislocki and Dempsey, 1955).

Other modern techniques in experimental biology have contributed important evidence on these questions, though of an indirect kind. In mesenchymatous tissues, on which Sedgwick laid stress, the relationships between adjacent cells have been strikingly elucidated by studies in tissue culture. When living fibroblastic cultures are examined by sufficiently sensitive optical methods a striking feature is the arrangement of fine filaments which stretch between the adjacent cells (Strangeways and Canti, 1927). These may vary in form with the state of extension of the cell. Although there is thus little doubt that mesenchymatous cells are interconnected by processes of the cell-membrane, yet equally cogent evidence from further experiments on tissue cultures demonstrates that these cells are nevertheless separate protoplasmic units.

Chambers and Fell in 1931 described the results they had obtained from the insertion of micro-needles into the cells of such cultures. Puncture of the nucleus was followed by degenerative changes within the cell after a few minutes; only where a cell had two nuclei was it possible to penetrate a nuclear membrane without causing the death of a cell. Wherever degeneration of a cell was brought about by this means, it rounded into an amorphous coagulated mass, without provoking any changes whatever in adjacent cells. Even in a sheet of intestinal or retinal epithelium in culture, in which cell-boundaries are virtually invisible during life, these authors observed that 'When one cell . . . was fatally damaged it shrank away from its neighbours except possibly at a few points, and degenerated, whilst the surrounding cells remained totally unaffected'.

Sedgwick's third instance of a process in animal embryology which he claimed was at variance with cellular theory was in the embryology of the nervous system. The development of our understanding of the manner in which the nerve fibre is formed is an important and particularly interesting chapter in the history of cellular theory. In the latter half of the nineteenth century there were two views on this question; either that a nerve fibre is an outgrowth from a single cell (Bidder and Kupffer, 1857; His, 1886) which with its fibres constitutes a functional cellular unit, termed by Waldeyer a 'Neurone',[1] or alternatively that nerve fibres differentiate within a continuous protoplasmic meshwork present from very early states of development (Hensen, 1864). The latter hypothesis is generally known as the 'protoplasmic bridge' theory; naturally Sedgwick subscribed to it, as later did his pupil Graham Kerr (1902, 1904).

The decisive demonstration of the truth of the rival neurone theory was largely the work of one man, Ross Harrison, who founded the technique of tissue culture with this purpose in mind. The production of protoplasmic fibres from neuroblastic cells isolated in a hanging-drop preparation and the outgrowth of these fibres into a fibrin clot where no pre-formed protoplasmic bridges could be postulated was capable of interpretation only in one way. In the developed and functional nervous system the elucidation of the neurone as a physiological unit has been one of the main themes in modern neurology, which has now extended widely throughout the animal kingdom. Even among the Coelenterates, where the nervous system was regarded even until recent years as a 'nerve net', the existence of synaptic junctions between contiguous neurones has now been fully demonstrated (Pantin, 1952).

Late nineteenth-century criticism of the cell-theory was not, however, confined to the questions of the reality of cell-boundaries in the Metazoa. In 1893 the distinguished American biologist C. O. Whitman contributed to the series of 'Biological Lectures' published by the Marine Laboratory at Wood's Hole an article with the title 'The Inadequacy of the Cell Theory of Development'. Whitman laid most stress on the comparison of a Protozoon individual with the whole body of a Metazoon.

[1] The form 'Neuron' was first used by B. G. Wilder in 1884.

The phenomenon of the regeneration of lost parts, he instanced, is essentially similar in the Protozoon *Stentor* and in the Metazoon *Hydra.* Furthermore, details of structure in the more complex Protozoa often exactly parallel those of Metazoa, and the presence of cell-boundaries in the latter does not affect the comparison. Thus in *Stentor* the arrangement of the characteristic fringe of large fused cilia and membranellae round the mouth is precisely the same as those of the 'corner cells' of the lamellibranch Mollusc *Cyclas.*

This important question of the individuality of the Protozoon, of the Metazoon body and of its relationship to the cell-theory was argued at greater length by Clifford Dobell in 1911. Dobell was a pupil of Sedgwick and had an equal command of his teacher's forensic style in scientific argument. Dobell's standpoint is clearly illustrated in quotations which are found early in his paper:

> The evolution theory and the cell theory, formulated as they were in the middle of last century, have had a paralysing effect upon the study of the Protista. These theories have forced men to see the Protista from an entirely subjective point of view. . . . So long as the Protista are 'primitive unicellular organisms', so long will their biological significance remain unrecognised.

And again:

> Amoeba is an entire organism in just the same sense that Man is an entire organism. As far as the concept 'individual' can be analysed . . . it is clear that a protist is no more homologous with one cell in a metazoon than it is with one organ (e.g. the brain or liver) of the latter. Only the cytologist blinded by what he sees through the microscope could ever believe in such a preposterous proposition.

The author then relates the history of the idea of a Protozoon as a single cell. This began quite early in the history of the cell-theory, mainly with von Siebold. Dobell instances the difficulties into which this notion has led a number of authors, mainly over multinucleate Protozoa. His own position was that Protozoa are by no means simple in structure or to be regarded as 'lower' organisms. They are neither uni- or multi-cellular, but non-cellular; they thus constitute an exception to the generalisation that the body of a living organism is divided into cells. This point of view has for long now been an accepted part of zoological teaching. By this time a great deal more has been learnt concerning the elaboration of structure within the more complex Protozoa, and within recent years the application of the electron microscope to this field has revealed some extraordinary examples of protoplasmic organisation in the absence of cellular compartments (e.g. Fauré-Fremiet, E., *et al*; 1956).

Dobell, however, did not confine his arguments to the Protozoa. He also asserted that the fertilised egg of a Metazoon is an individual organism and not a single cell, for the obvious reason that the egg will itself develop into a whole organism. When a zygote divides into two blastomeres the whole remains an individual, now, however, composed of two cells. Here, albeit, we meet a difficulty. If these two cells become parted, either by the agency of an experimenter, as with Driesch (1892), or spontaneously, as can occur when identical twins develop from a single Mammalian egg, the separated cell with no apparent further change then at once advances in status to an organism, for it proceeds to develop into a complete individual. Here perhaps is the weakest aspect of Dobell's viewpoint, though everywhere else it has much more to recommend it than has any other.

In general it must be admitted that nowhere in biology is there a theory wholly without objection or a generalisation to which no exception has to be admitted. It is this feeling of the elusiveness of the absolute which still remains after one has read a paper such as Dobell's, argued however forcibly in triumphantly refuting some seemingly outworn hypothesis.

In the growth of a branch of science a theory may fulfil several purposes. In the first place it groups together hitherto isolated observations and indicates a common link between them. It may suggest, furthermore, where to seek for fresh facts relevant to the whole field of learning. In all this it exerts a subjective function in arousing the interest of enquirers. It might well be claimed that the final importance of a theory in the history of any science lies in the extent to which it has provoked thought and stimulated further enquiry. In retrospect it may be clear that the subsequent effort of research was largely directed into channels which in the end were found to be mainly irrelevant to the original group of problems. But the new knowledge is finally won, however tortuous the approach. It would not be difficult to find examples to illustrate this viewpoint from any of the classical theories which served as the scaffolding for the fabric of biology in the nineteenth century, and instances are ready to hand within our present topic.

In 1895 Max Verworn published his *General Physiology*. This work was based entirely on the thesis which in the same year Sedgwick so vigorously rejected. For Verworn the cell was 'the structural element of the living body, the elementary organism in which the vital processes have their seat' (1899, p. 48). His declared object was to found a physiology of the cell. He proceeded to discuss how this might be achieved; he admitted that observations on individual cells within a tissue were then seldom possible under physiological conditions, although this aim had occasionally been effected, as in Heidenhain's studies on the salivary glands (1883). Verworn, however, recalled the emphasis which Johannes Müller had laid on the importance of the comparative method of physiological enquiry, which

> demonstrates one fact of fundamental importance, namely that the elementary vital phenomena belong to every cell, whether it be from a tissue of the higher animals, the lower animals, the plants, or free-living and independent unicellular organisms. Every one of these cells exhibits in its individual form general vital phenomena. Realising this, it is only necessary for the investigator to select from the variety of species the objects best fitted for each special research. . . . It is no longer necessary for him to cling to the tissue-cells of the higher vertebrates alone, which can be employed for microscopic experiments alive and under normal vital conditions only in rare and exceptional cases. . . . Much more favourable in this respect are the tissue-cells of many invertebrates. . . . But the free-living unicellular organisms,

the *Protista,* appear to be the most favourable objects for cell-physiological purposes. They seem to have been created by nature for the physiologists, for, besides their great capacity of resistance, of all living things they have the invaluable advantage of standing nearest to the first and simplest forms of life; hence they show in the simplest and most primitive form many vital phenomena that by special adaptation have developed to great complexity in the cells of the cell community. [Ibid., p. 51.]

In his choice of the Protozoa as material for physiological studies, whether on grounds of convenience or because of their supposedly primitive nature, Verworn thus tacitly accepts a homology between a Protozoan individual and the Metazoan cell. Elsewhere in his book he was able to refer to a considerable body of physiological studies on Protozoa in which the same assumption can be traced. Thus, for instance, from the mid-80's several authors began to attack the question of the function of the nucleus by experimenting on the effect of dividing Protozoa into fragments. Micro-dissection of this kind had begun even in the previous century (Eichhorn, 1783), but in the light of the discoveries in nuclear cytology of the 1880's such experiments were again undertaken with a fresh purpose. Both Nussbaum (1884) and Grüber (1885) showed that the ability of a severed portion of a Ciliate to regenerate into a smaller but complete individual depended on the presence within the fragment of nuclear material.[1] Balbiani in 1887 (published 1892) studied in considerable detail the effects of merotomy on the bodily functions of Ciliates, while Hofer (1890) found that when a large amoeba was divided normal pseudopodial movement continued in the portion deprived of the nucleus for only 15–20 minutes.[2]

Again Verworn's treatment of the topic of 'stimuli and their

[1] Actually in most cases part of the meganucleus. The distinction between micro- and meganucleus in the Ciliata was not established until 1888 by R. Hertwig.

[2] Verworn himself did not admit that such experiments demonstrated the supremacy of the nucleus. This, he asserted, could only be admitted if the nucleus by itself was able to regenerate a cytoplasm around it. In the radiolarian *Thalassicola* he had found that an isolated nucleus, 'even when it is protected from all injury', fails to do this.

actions' deals largely with the effects of various agencies on living Protozoa. He draws extensively upon his own experiments on marine Radiolaria, in the exquisitely fine pseudopodia of which he sees an analogy with the nerve fibres of the Metazoa.

Radl (1930, p. 241) tells us that 'physiologists gave the book a cool reception'. A review of the *General Physiology* in the pages of *Nature* (1895, p. 529) written by Michael Foster was not unsympathetic in tone, yet clearly showed how the subject was then firmly set on a different path:

> It is not for me, who in my rash youth had wild dreams of building up a new physiology by beginning with the study of the amoeba and working upwards, to say one word against the experimental investigation of the lower forms of life. But experience and reflection have shown me that, after all, the physiological world is wise in spending its strength on the study of the higher animals. Taking . . . as an instance, the molecular processes which give rise to the movement of animals, and which appear under such forms as that of amoeboid movement and that of the contraction of a striated muscle, I venture to think that the very apparent simplicity of the former is an obstacle to our getting a real grasp of its inner nature, and that by our studies of the complex muscle we are drawing nearer to such a grasp than we could ever have done by observations confined to the phenomena of the amoeba itself. And so in many other instances, the study of the lower forms of life is, in reality, more difficult than that of the higher forms, and the latter naturally comes first.

Foster's choice of protoplasmic movement and contraction was the perfect illustration of his thesis, in the subsequent history of which he has been entirely justified. All that is yet known about amoeboid or ciliary movement has first been learnt from the study of striated muscle and has later been applied thereto. Yet only in recent years has it been possible to overcome the difficulty of approach to the single highly differentiated cell, of which Verworn was so aware. The studies of such workers as A. F. Huxley and his colleagues (1954) on single muscle fibres have marked this final stage of achievement.

The development of the cellular physiology towards which in his own day Verworn looked forward is a major feature of the biology of the present time, in which it is of great interest to observe that a detached attitude towards the idea of the cell as the unit of life has been maintained. A milestone in the development of cellular biology in this century was the publication of the *Experimental Cytology* of James Gray in 1931. In this book there is early revealed the author's attitude to this question, one which still shows something of the influence of Sedgwick and Dobell:

> Cellular structure is, teleologically, a mechanical necessity for life in large and varied forms. From this point of view, the cell as a unit of life is both unnecessary and unsatisfactory—it is merely the unit of mechanical stability. The real unit of life must be of a protoplasmic nature irrespective of whether it is subdivided to form a mechanically stable system or not: in other words, cellular structure is not by itself of primary significance. If we take a biologically heterogeneous system of growing protoplasm and proceed to a process of internal subdivision there may come a time when each phase of the system will be separated from the others by cell walls. At this stage each cell will represent a natural protoplasmic unit—but before this stage is reached, the only real unit available is one which is expressed in terms independent of the process of sub-division. There can be little doubt that the most natural unit of life is the living organism, and when we find, in some cases, that its constituent cells are united by intercellular processes, it is impossible to admit the validity of the cell unit without further enquiry.
>
> On the other hand, there can be no doubt that the cell often forms a convenient physiological unit even if its individuality is not so fundamental as is sometimes supposed. Each living cell possesses, structurally, the essential machinery for independent existence; each cell normally has a nucleus and is chemically and physically in equilibrium with its environment by means of its surface membranes.

Still more recently, the various contributors to the volume entitled *Cytology and Cell Physiology* (ed. Bourne, 1942 and

1951) have found themselves under no necessity to make any reference whatever to this question.

Our present thesis that during the second half of the nineteenth century the concept of cellular individuality served as a primary stimulus to the development of microscopical biology can also be illustrated from the history of contemporary studies in embryology. Here again we may see that cellular theory provided a framework within which the subject could develop, but from which a large measure of detachment was ultimately achieved.

The question which may be said to underlie much of later nineteenth-century study of the early stages of development is that of the relationship between cleavage and differentiation. In the 1870's detailed embryological studies in various phyla of the animal kingdom began to reveal instances where during the cleavage of the egg a gradual unfolding of a plan of development was discernible within a fixed pattern of cell divisions. The arrangement of the cells even in the early embryo was thus unmistakeably related to basic morphological features of the differentiated organism; and the production of the individual cellular units was apparently the means through which there emerged a total organic pattern.

The earliest examples of this kind to be discovered were found among Annelids and Molluscs, the early development of both of which was later shown to conform to a common plan. Kowalevsky in 1871 found that in the earthworm the first mesodermal cells are arranged in longitudinal bands, at the growing apex of which is a large cell, from which all the others have been cut off by a succession of unequal divisions. Such cells later received the name of 'teloblasts' (Wilson, 1887). Whitman in 1878 found a similar mode of development in the leech and traced the whole of the nerve cord to an origin from a single pair of cells. In the following year Rabl (1879) followed the earliest pattern of cleavages in the pond-snail *Planorbis* and described that regular alternation in the inclination of the planes of division by which successive 'quartets' of cells become arranged in a characteristic pattern. Rabl traced the entire ectoderm of the developing embryo to three groups of blastomeres which are formed in succession.

Among chordate embryos there were discovered instances in which the orientation of the first plane of cleavage approximately coincided with the future median plane of the embryo. As far back as 1854, George Newport, in that remarkable study of fertilisation in the frog, of which mention was made in an earlier page (60), had even grasped the fact that the first cleavage furrow not only lay in the future median plane of the embryo but was determined by the point of entry of the sperm. This he had proved by applying spermatic fluid on the point of a pin to the ripe egg. Thirty years later van Beneden and Julin (1884) observed that in the eggs of ascidians also the fertilised egg divided first in such a way as to separate the future left and right halves of the embryo. In the same year Wilhelm Roux, apparently unaware that he was but confirming an observation first made when he was four years old, published a study on the first cleavage planes of the frog's egg (Roux, 1884).

By that time, a few years after Flemming's synthesis, the main facts concerning the division of the nucleus were generally familiar to biologists. In the previous year Roux (1883) had published a commentary on nuclear division which he attempted to relate to his observations on the cleavage planes in the egg of the frog. He began by arguing that the facts of karyokinesis were only explicable on the assumption that chromatin is not a uniform and homogeneous substance but differs qualitatively in different regions of the nucleus, and that the collection of the chromatin into a threadwork and its accurate division into two halves is meaningless unless the substance in different regions represents separate qualities to be divided and distributed according to some definite plan. Otherwise, he cogently added, a simple and direct division of the nucleus would be equally efficacious. Here, then, was a brilliant and intuitive anticipation of a principle well over half a century before its ultimate proof. To us, the notion implies the complete equality of the daughter nuclei which are formed at each division. Roux, however, in order to link embryonic differentiation with nuclear division, vitiated his whole argument by postulating that karyokinesis could result in either a qualitative or a quantitative distribution of nuclear qualities. The earliest divisions of the zygote nucleus, he believed, were of the former kind. The first cleavage of the frog's egg separates materials destined to form the left and right halves of the embryo. Roux believed that the distinction between the first two blastomeres resided in the first instance within the chromatin, appropriate kinds of which were distributed to either end of the first mitotic figure across the future median plane of the embryo. The test of experiment when applied to this question at first seemed to confirm this view. Roux's well-known experiments (Roux, 1888) in which he cauterised one blastomere in the frog's egg after the first cleavage apparently proved that each blastomere at this stage is 'determined', to use Roux's own term, to form exactly and inevitably a half embryo. Moreover, in the previous year Chabry (1887) had obtained essentially similar results in the tunicate *Ascidiella* by puncturing one of the first pair of blastomeres.

However, even at that time other experiments had shown that it is possible to distort the pattern of the early cleavage planes in the frog's egg, and so disturb the normal distribution of cytoplasmic material and nuclei. Yet under these circumstances a normal embryo can still be obtained. This experiment was first performed in 1884 by Pflüger, who compressed the egg between parallel plates during the first three cleavages, none of which occurred in the plane of these compressing surfaces.

In 1892 Hans Driesch made the astonishing discovery that if the first two blastomeres of the sea-urchin egg were separated, then each would develop into a whole larva, differing only in size from the normal *pluteus*. The contrast between the results of Roux and of Driesch led to the admission of a distinction between two general types of cleavage, named by Conklin (1897) 'determinate' and 'indeterminate'. Subsequent researches showed that no absolute distinction exists between the two types. Thus the separation of blastomeres in the sea-urchin after the third cleavage no longer always results in the development of perfect larvae from each isolated cell (Driesch, 1900). Moreover, the course of Roux's original experiment on cauterising the frog's egg during cleavage is influenced by the necrotic cytoplasm of the affected blastomere.

It is now clear that cleavage and cytoplasmic differentiation are separate and distinct processes with no necessary causal connection between them. They both belong to the earliest phases of embryonic development, but the relative times at which each begins vary from one organism to another. A remarkable instance of the artificial disengagement of the two processes was provided by F. R. Lillie (1902), who, after treating the fertilised egg of the Annelid *Chaetopteris* with potassium chloride, found that cilia developed at the surface of the egg in the complete absence of segmentation. The significance of this result was enhanced by the fact that the cleavage of Annelids is highly determinate.

Radl (1930), in a highly unsympathetic account of the history of the cell doctrine, now largely outdated in treatment, dismisses the long series of researches inspired by the cell-theory in these words:

> It is as if the clue to all living problems were hidden in the cell, as if the microscope could disclose to us all the unknown springs of being [p. 231].

It would now, one hopes, be agreed that no single instrument can guide us to these mysterious headwaters, or that they are approachable through one single order of magnitude alone. Yet we can still learn from the history of such attempts by past explorers, even though their initial assumptions have now been superseded. It would be legitimate, though hardly profitable, to criticise the manner of Columbus's discovery of America.

LITERATURE FOR CHAPTER SIX

ACKERKNECHT, E. H. (1953). Rudolf Virchow, Doctor, Statesman, Anthropologist. Univ. Wisconsin Press.

BAKER, J. (1952). Quart. J. Micro. Sci. 93, 157.

BALBIANI, E. G. (1892). Ann. Micrograph. 4, 369.

DE BEER, G. R. (1940). Embryos and Ancestors. Oxford.

VAN BENEDEN, E., AND JULIN, C. (1884). Arch. de Biologie. 5, 111.

BIDDER, F. H., AND KUPFFER, C. (1857). Untersuchungen über die Textur des Ruckenmarks und die Entwickelung seiner Formelemente. Leipzig.

BOURNE, G. C. (1896). Quart. J. Micro. Soc. 38, 137.

BOURNE, G. (ed.) (1942). Cytology and Cell Physiology. Oxford.

BOYD, J. D., AND HUGHES, A. F. (1954). J. Anat. London. 88, 356.

CHABRY, L. (1887). Journ. Anat. et Physiol. 23, 167.

CHAMBERS, R., AND FELL, H. B. (1931). Proc. Roy. Soc. (B) London. 109, 380.

CONKLIN, E. G. (1897). J. Morphol. 13, 1.

DOBELL, C. (1911). Archiv. f. Protistenkunde. 23, 269.

DRIESCH, H. (1892). Zeit. f. wiss. Zool. 53, 160.

DRIESCH, H. (1900). Arch. Entw.-Mech. 10, 361.

EICHHORN, J. E. (1783). Wasserthiere . . . Zugabe dazu. Danzig.

FAURÉ-FREMIET, E., ROUILLER, C., AND GAUCHERY, M. (1956). Arch. d'anat. Micro. et de Morph. exp. 45, 139.

FOSTER, M. (1895). Nature. 51, 529.

GARDINER, W. (1884). Arb. bot. Inst. Wurzburg. 3, 52.

GRAY, J. (1931). A Text-Book of Experimental Cytology. Cambridge.

GRÜBER, A. (1885). Biol. Centrbl. 4, 717.

HAECKEL, E. (1866). Generelle Morphologie der Organismen. I allgemeine anatomie der arganismen. Berlin.

HARRISON, R. (1907). Anat. Rec. 1, 116.

HEIDENHAIN, R. P. H. (1883). In Hermann, L., Handbuch d. Physiologie. 5, 1. Leipzig.

HEITZMANN, C. (1883). Mikroskopische Morphologie d. Thierkorpers . . . Vienna.

HENSEN, V. (1864). Arch. f. path. Anat. u. Physiol., p. 176.

HERTWIG, R. (1888). München Ges. Morphol. u. Physiol. Sber. 3, 127.

HIRSCHFELD, E. (1929). Jahrb. d. Inst. fur. geschichte des Med. Leipzig. 2, 106.

HIS, W. (1886). Arch. Sci. Phys. Nat. 16, 344.

HOFER, B. (1890). Jen. Zeit. 24, 105.

HUXLEY, A. F., AND NIEDERGERKE, R. (1954). Nature. 173, 971.

HUXLEY, A. F., AND TAYLOR, R. E. (1954). Ibid. 176, 1068.

HUXLEY, L. (1900). Life and Letters of T. H. Huxley. Vol. 1. London.

HUXLEY, T. H. (1853). Brit. and For. Medico.-chir. Review. 12, 285.

KERR, G. (1902). Quart. J. Micro. Sci. N.S. 46, 417.

KERR, G. (1904). Trans. Roy. Soc. Edinburgh 41 (1), 119.

KOWALEVSKY, A. (1871). Mem. de l'acad. imp. des Sci. St. Petersburg. Ser. 7, No. 16.

LILLIE, F. R. (1902). Arch. f. Entw. Mech. 14, 477.

MANTON, S. M. (1949). Philos. Trans. (B) Roy. Soc. London, p. 529.

NEWPORT, G. (1854). Ibid. p. 229.

NUSSBAUM, M. (1884). Bonn. Niederrhein. Ges. Sber., p. 259.

PANTIN, C. F. A. (1952). Proc. Roy. Soc. (B) London. 140, 147.

PFLÜGER, E. (1884). Arch. ges. Physiol. 34, 607.

RABL, C. (1879). Morph. Jahrb. 5, 562.

RADL, E. (1930). Geschichte der biologischen Theorien . . . Trans. E. J. Hatfield. Oxford.

REICHERT, K. B. (1855). Arch. f. Anat. Physiol. wiss. Med., p. 1.

ROBERTS, M. (1938). Biopolitics: an essay on the physiology, pathology and politics of the social and somatic organism. London.

ROUX, W. (1883). Ueber die Bedeutung der Kerntheilsfiguren. Leipzig.

ROUX, W. (1884). Deutsch. Natf. Tagebl., p. 330.

ROUX, W. (1888). Schles Ges. Jber. Breslau, p. 267.

SCHLEIDEN, M. J. (1838). Beitrage zur Phytogenesis. Arch. f. Anat. und Physiol. Trans. H. Smith, (1847). Sydenham. Soc., London.

SEDGWICK, A. (1885). Quart. J. Micro. Sci. 25, 449.

SEDGWICK, A. (1894). Ibid. 36, 35.

SEDGWICK, A. (1895). Ibid. 37, 87.

SPENCER, H. (1893). Contemporary Review. 63, 439.

STRANGEWAYS, T. S., AND CANTI, R. G. (1927). Quart. J. Micro. Soc. 71, 1.

STRASBURGER, E. (1901). Jahrb. wiss. Bot. 36, 493.

TEMKIN, O. (1949). 'Metaphors of Human Biology', in Science and Civilization. Ed. R. C. Stauffer. Madison, Wisconsin.

VERWORN, M. (1895). Allgemeine Physiologie. Jena. Trans. F. S. Lee (1889), from 2nd German ed. London.

WHITMAN, C. O. (1878). Quart. J. Micro. Soc. 18, 215.

WHITMAN, C. O. (1893). J. Morphol. 8, 639.

WILSON, E. B. (1887). J. Morphol. 1, 183.

WISLOCKI, G. B., AND DEMPSEY, E. W. (1955). Anat. Rec. 123, 133.

Explanation of Plates

PLATE I HISTORIC MICROSCOPES

Fig. A. Simple microscope of Leeuwenhoek. Reproduced from Disney Hill and Watson Baker (1928) p. 160. The original was made before 1673.

Fig. B. Hooke's microscope. From the Micrographia (1665); arranged for indirect lighting. Some description of the microscope is given in the Preface:

'The tube being for the most part not above six or seven inches long, though, by reason it had four Drawers, it could very much be lengthened, as occasion required; this was contrived with three glasses . . .

the stand also is described:

'Upon one side of a round Pedestal AB in the sixth figure of the first *scheme*, was fixt a small Pillar CC, on this was fitted a small iron arm D which could be moved up and down, and fixed in any part of the Pillar, by means of a small screw E; on the end of this arm was a small Ball fitted into a kind of socket F, made in the side of the Brass ring B, through which the small end of the Tube was screw'd, by means of which contrivance I could place and fix the Tube in what posture I desired (which for many observations was exceedingly necessary) and adjusten it most exactly to any object.'

Fig. C. Reflecting microscope of G. B. Amici (1820), reproduced from Disney, Hill and Watson Baker (1928) p. 234, who thus described it:

'The system employed with this microscope was to reflect by means of a prism the image of the object on to a concave mirror, which in turn formed the image which was examined and magnified by the eyepiece.'

Fig. D. Compound achromatic microscope of Ploessl, first introduced about 1830. The figure is from Berres (1837).

PLATES II—VI FIGURES OF SPERMATOZOA TAKEN FROM VARIOUS AUTHORS

PLATE II

A. From Leeuwenhoek (1679); ×200 approx. Figs. 1–4 Human; figs. 5–8 of the Dog.

B. From Ledermüller (1758); ×400 approx. Fig. 58—Human; fig. 61—of a tortoise; fig. 62—of the perch; fig. 63—of the frog; fig. 64—of a snake.

PLATE III

From Prévost and Dumas (1821); ×3000. Fig. 1—Of the guinea pig; fig. 2—of the white mouse; fig. 3—of the hedgehog; fig. 4—of the horse; fig. 5—of the cat; fig. 6—of the ram; fig. 7—of the dog.

PLATE IV

A. From Dujardin (1837); ×800–1000. Fig. 6—Human; fig. 8—of the guinea pig (fig. 8*a* is ×300); fig. 9—of the mouse.

B. From Wagner (1837); ×800.

C. From Valentin (1839); of the bear. [From the legend: (*c*) mouth; (*d*) anus].

D. From Pouchet (1847). Fig. 3 from his Plate XV, of the rabbit; fig. 9 from his Plate XIX, human, showing "les vestiges d'organisation".

E. From Gerber (1842); of the guinea pig. [From the legend: (*b*) Internal vesiculi (probably botyroidal stomach). (*c*) Two globular organs. (*d*) Oral aperture. (*e*) Genital and anal orifice.]

PLATE V

From Donné (1845) ×400. (*a*) of the frog; (*b*) of a bat; (*c*) human; (*d*) of the mouse.

PLATE VI

A. Spermatozoon of mouse from Schweigger-Seidel (1865), who recognised the existence of the middle piece.

B. Part of mouse sperm, from Leydig (1883). Oblique "cross-striation" in middle piece.

C. Testicular sperms of mouse from von Brunn (1884), showing aggregations of granules to form the "bands" round the middle piece.

D. Rat spermatozoa, after slight maceration, from Jensen (1887). In (*i*), by slight unravelling, the true nature of the spiral filament of the middle piece has been shown. In (*ii*) the constituent filaments of the tail are partly separated.

E. Rat spermatozoa from Retzius (1909).

PLATE VII

THE BEGINNINGS OF PLANT HISTOLOGY

A. Robert Hooke's drawings of the "Schematisme or Texture of Cork", from Scheme XI of the Micrographia (1665).

B. Tab. 28 from Nehemiah Grew's "Anatomy of Plants" (1682); of an "Elm Branch cut transversely."

C. Part of Tab. 40 from the same; of a "Sumach Branch cut transversely" as magnified by Baker (1952), to illustrate Grew's mistaken views of the nature of cells and cell walls.

PLATE VIII

EARLY OBSERVATIONS ON THE STRUCTURE AND DIVISION OF CELLS

A. Figures from Leeuwenhoek (1702) of the red blood corpuscles of fishes. In these corpuscles he saw "a little clear sort of a light in the middle". These are the earliest drawings of nucleated cells.

B. Group of epithelial cells in the slime of an eel, from Fontana (1781). The nucleus is described as "un corps oviforme, pourvu d'une tache en son mileau", probably the nucleolus.

C. Nerve cell drawn by Valentin (1836). Nucleus and Nucleolus are represented, through the nuclear membrane is incomplete.

D. Nerve cell of a Crayfish, from Remak (1844) showing neurofibrillae in the cytoplasm, orientated in parallel lines.

E. Multinucleate tumor cells from Virchow (1851) supposed by him to represent the endogenous formation of new cells.

F, G. Figures which illustrate Schleiden's views on the formation of cells from the "Beiträge", (1838); taken from the Sydenham Soc. Edition of 1847, but with contrast reversed.

F. purports to illustrate free cell formation in the embryo sac of *Chamaedorea*.

G. shows the nucleus (or cytoblast) embedded in the cell wall.

H. Cartilage cells from an amphibian larva from Schwann (1839), taken from the Sydenham Soc. edition of 1847, purporting to show free cell formation in the intercellular matrix.

I. Fission in the Diatom *Synedra* from Trembley, as reproduced by Baker (1951).

J. The primordial utricle of the mature plant cell, as depicted by Von Mohl (1845). The protoplasm lining the interior of the cell wall becomes evident by shrinkage after fixation in alcohol.

PLATE IX

EARLY OBSERVATIONS ON THE CELL IN DEVELOPMENT

A. Drawing of a female *Rhabditis* from Roffredi (1775), showing nuclei within oocytes, eggs, and embryos.

B. Stages in the cleavage of the Frog's egg, from Prévost and Dumas (1824).

C. Cleavage within the blastoderm of *Sepia*, from Kölliker (1844). In I–IV, the nucleus (*b*) is labelled "Embryonalzelle sammt Kern" and in the later figures the inner, separated blastomeres are "Furchungskugeln."

D. Figures showing the extrusion of the polar bodies in the fertilized egg of *Limax*, from Warneck (1850).

E. Some of Purkinje's original illustrations (1830) of the oocyte nucleus of the hen's egg, the "vesicle of Purkinje".

F. Two figures from Derbès (1847), of the sea-urchin egg; showing spermatozoa within the fertilization membrane, and astral rays within the cytoplasm.

G. Reichert's (1847) figure of the testis of the Nematode *Ascaris* showing asters within spermatocytes in meiosis.

H. Stages in the mitosis of erythroblasts in the chick embryo, from Remak (1858).

PLATES X AND XI

FIGURES OF NUCLEAR DIVISION AND FERTILIZATION

PLATE X

Figs. A, B and C are of nuclear division in the terminal cells of the staminal hairs of *Tradescantia*.

A. From Nägeli (1844).

B. From Hofmeister (1849).

C. From Strasburger (1880), which represent a series of observations on one cell during its division, at the following times in minutes after that of No. 38:(39) 60, (40) 85, (41) 115, (42) 130, (43) 135, (44) 140, (45) 150, (46) 155, (47) 175, (48) 215.

D. Selected figures from Hofmeister (1848) of pollen mother cells of *Tradescantia* in division (re-arranged).

E. Figures from Flemming (1882) of the division of a living cartilage cell from a Salamander larva.

F. Figures from Flemming (1882) of dividing epithelial cells from the same source. Fixed and stained.

G. Conjugation in *Paramoecium* illustrated by Balbiani (1861). The micronuclei (*b*) are shown both in prophase (*top left*) and elsewhere in metaphase, though the author at that time was mistaken in their interpretation.

PLATE XI

A. Figures of fertilization and development of the embryo in the orchid *Morio* from Amici (1847).

B. Free cell formation in the embryo sac of *Reseda odorata*, from Strasburger (1884). The right side of the figure where cell walls have been formed, is more advanced in development.

C. Figures from Hertwig (1878) of the formation of both polar bodies in *Asteracanthion*, by unequal cell division.

D. Fertilization by a single sperm in *Asterias glacialis* from Fol (1879). Two successive figures.

E. Fertilization and the formation of polar bodies in *Ascaris megalocephala bivalens* from O. Hertwig (1893). (*a*) Spermatozoon attached to egg. (*b*) Spermatozoon inside egg; anaphase of first maturation division. (*c*) Chromosomes in male pronucleus. First polar body given off. (*d*) Anaphase of second maturation division. (*e*) Contact between pronuclei; each with two chromosomes. (*f*) Metaphase of first cleavage division.

F. Figures of cleavage in *Ascaris* from Boveri (1888), as redrawn by Wilson, which demonstrate the continuity of the chromosomes in interphase. *Upper*: first telophase with nuclear membrane extended round ends of chromosomes. *Lower:* ensuing prophase.

PLATE XII

OBSERVATIONS ON NUCLEUS AND CYTOPLASM

A. Egg of *Ascaris megalocephala* in metaphase of first cleavage; from Van Beneden and Neyt (1887), showing astral centres.

B. A primary spermatocyte of *Brachystola magna* in early prophase, from Sutton (1902). The two views are at different levels of the same nucleus; between them are seen all eleven pairs of autosomes (*a–k*). The unpaired chromatin element 'x' is seen in both views.

C. The "quadrille of the centres" from Fol (1891).

D. Synapses of bivalents in the spermatogenesis of *Batrachoceps*, from Janssens (1905), as reproduced by Morgan (1919).

E. Mitochondria within fibroblastic cell in tissue culture from Lewis and Lewis (1914).

F. Part of giant chromosome from Salivary gland of *Chironomus*, from Carnoy (1884).

G. Camillo Golgi's original figure of the impregnated network within a Purkinje neurone of the barn owl (1898).

H. Two figures from Hardy (1899), showing the effect of fixatives in producing networks of different texture within gut cells of *Oniscus*. (*a*) Corrosive sublimate. (*b*) Osmic acid.

PLATE I

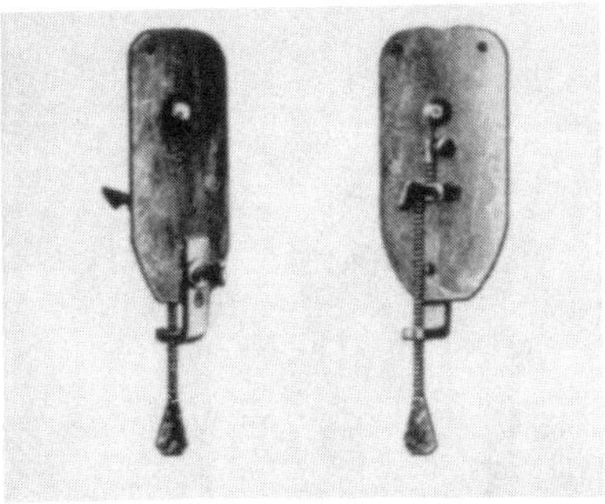

A

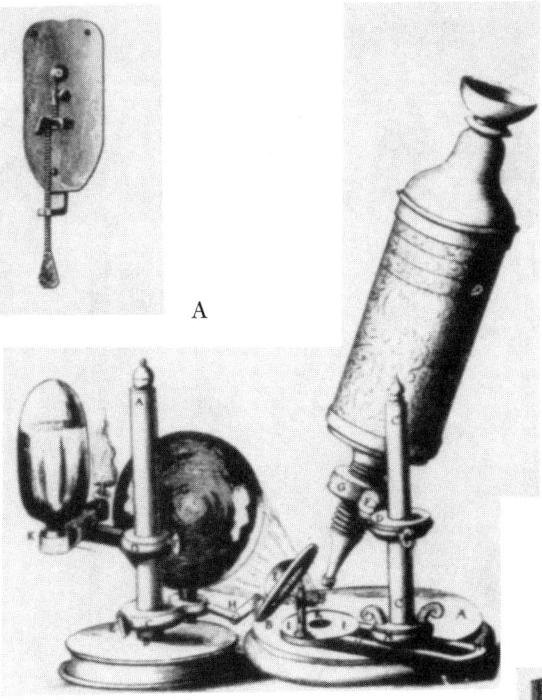

B

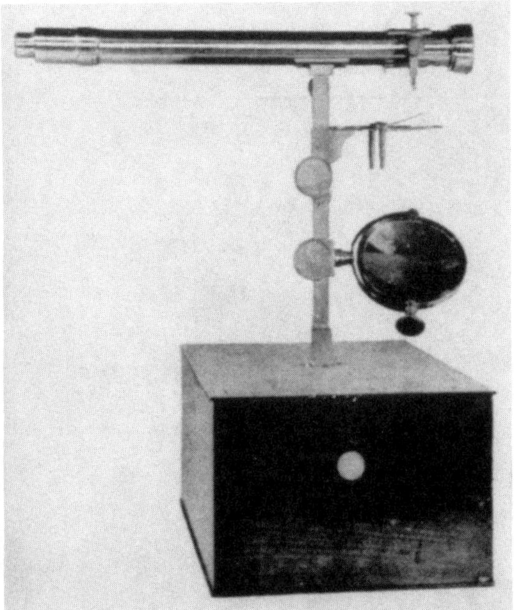

C

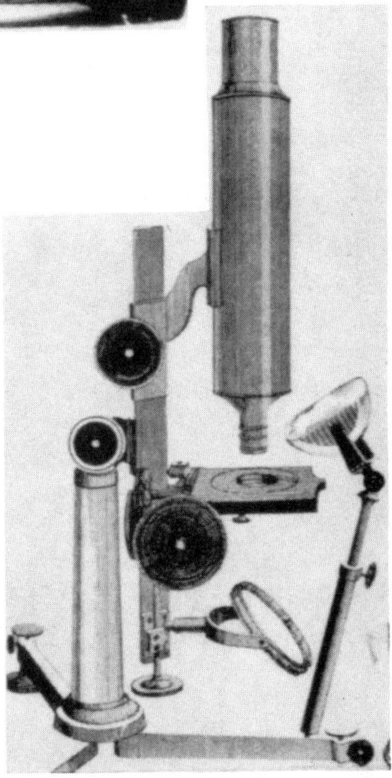

D

PLATE II

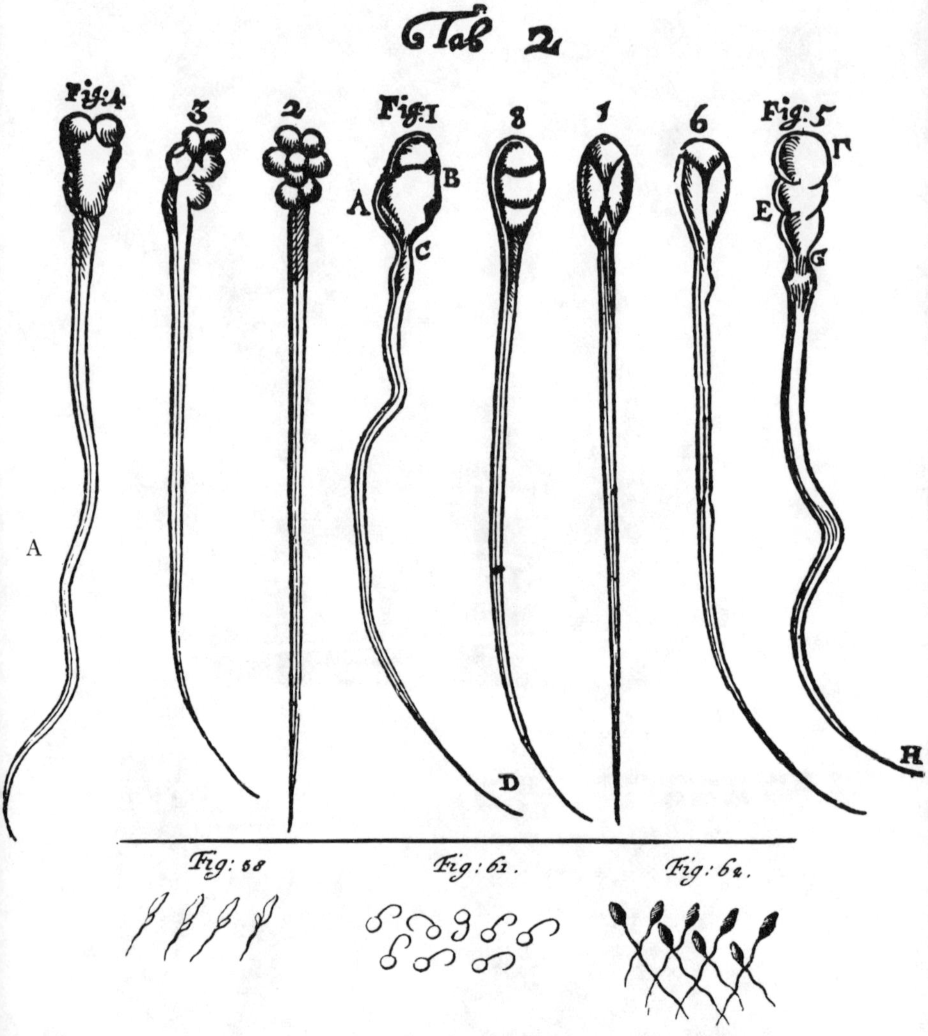
Tab 2
Fig:4
3
2
Fig:1
8
1
6
Fig:5
A
B
C
D
Γ
E
G
H

A

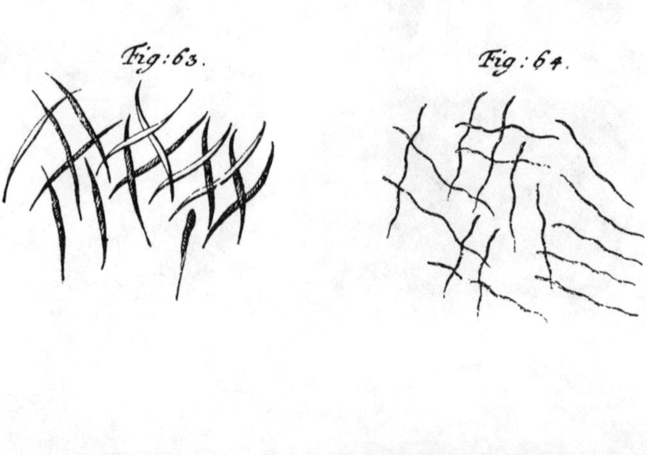
Fig: 68
Fig: 61.
Fig: 62.
Fig: 63.
Fig: 64.

B

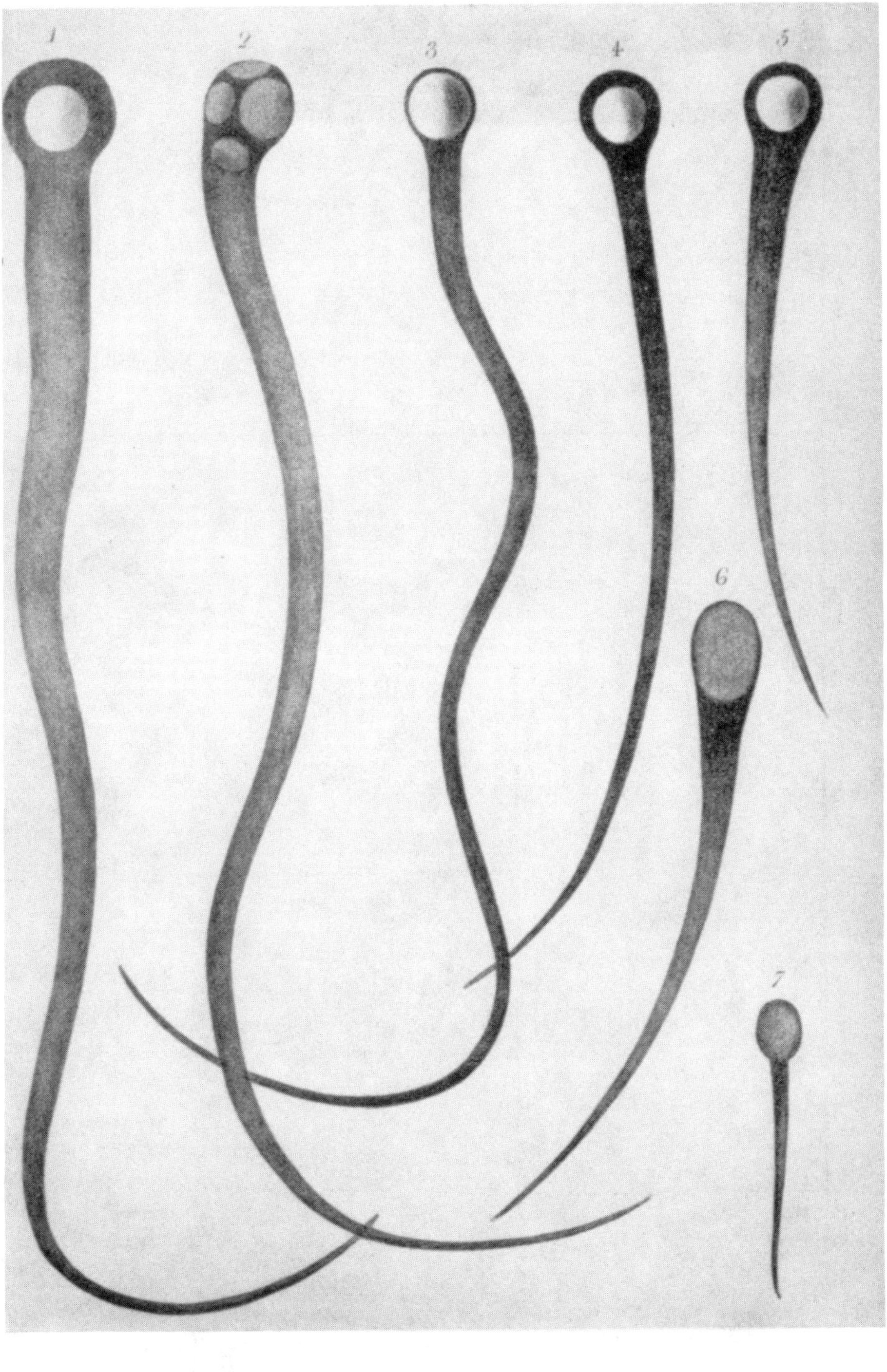
1
2
3
4
5
6
7

Plate IV

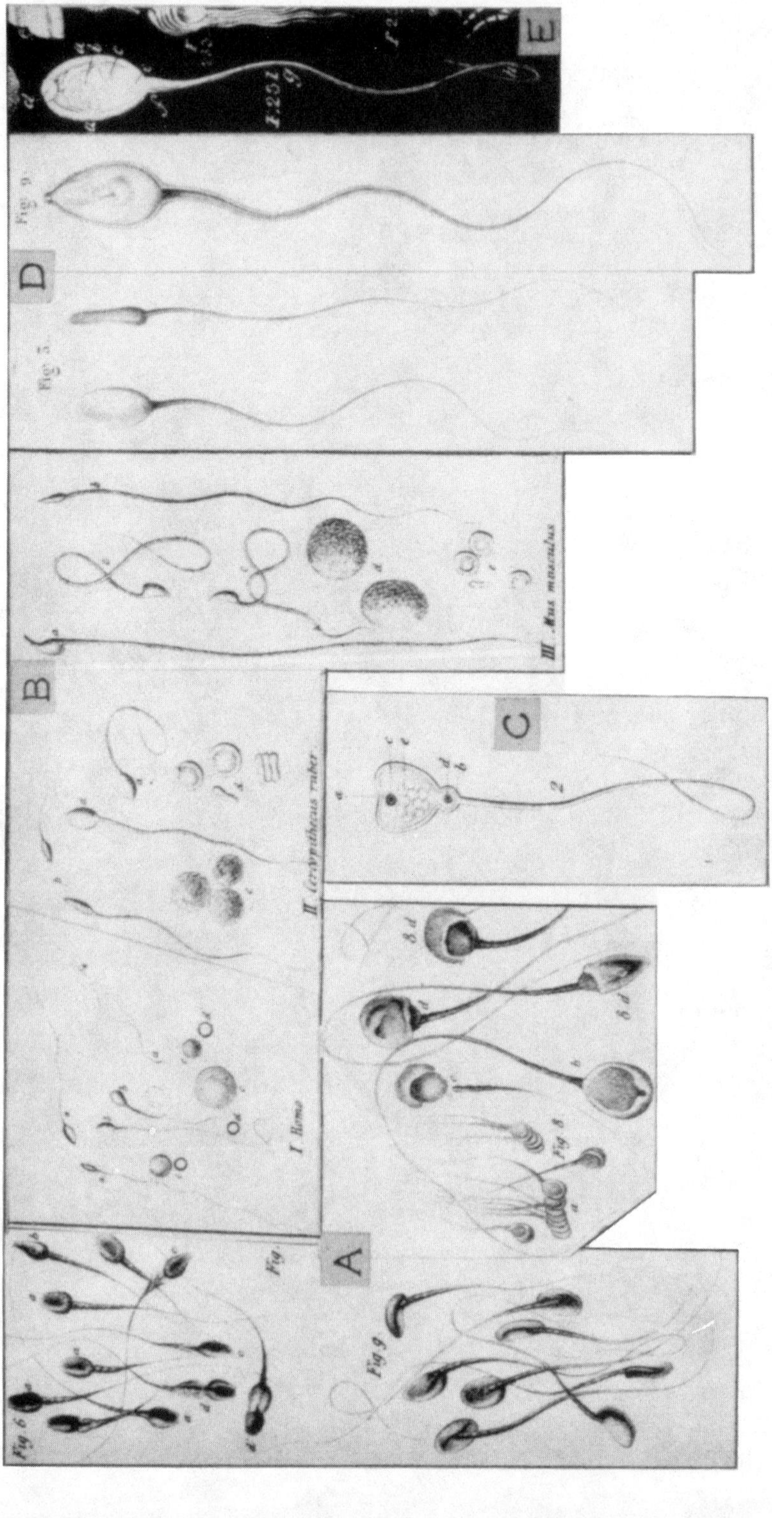
E
D
Fig 9
Fig 5
III Mus musculus
B
II Cercopithecus ruber
C
I Homo
Fig 8
A
Fig 6
Fig 9

PLATE V

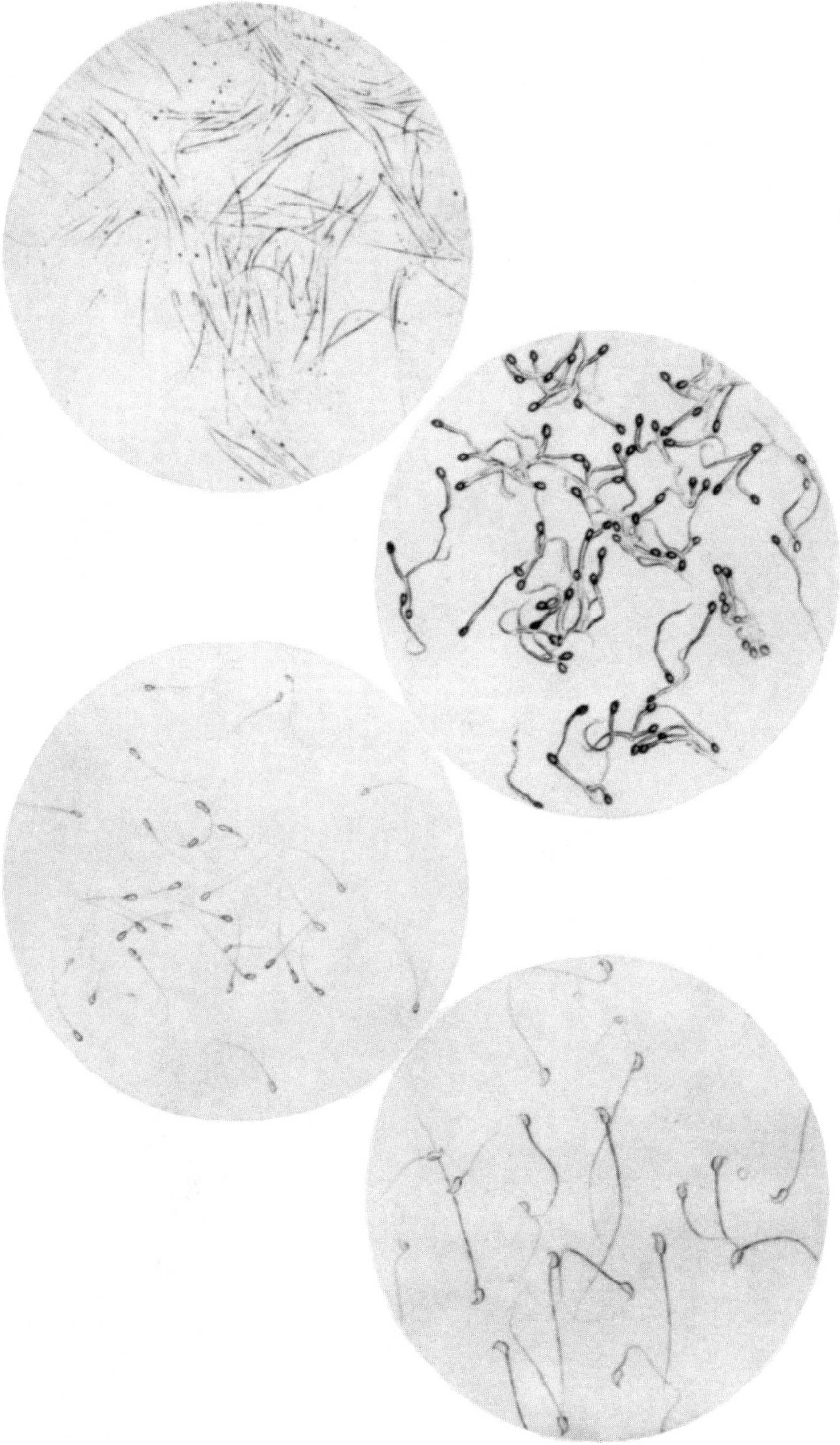

(a)

(b)

(c)

(d)

PLATE VI

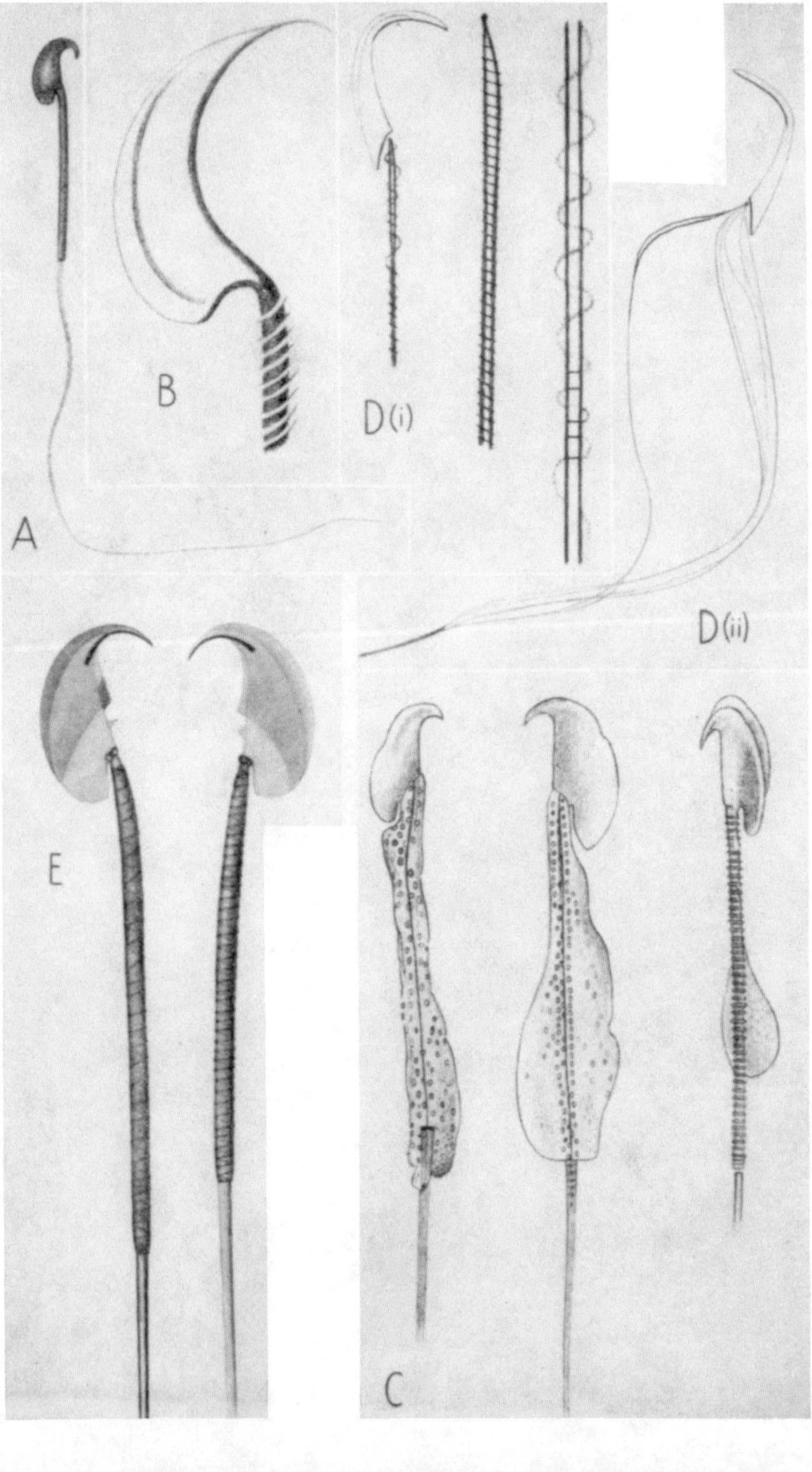

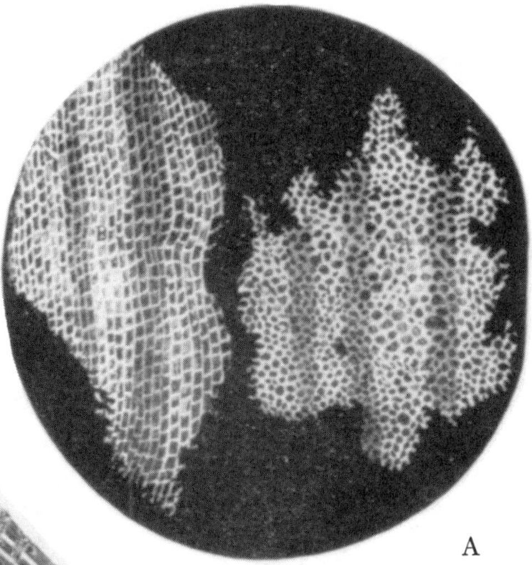

A

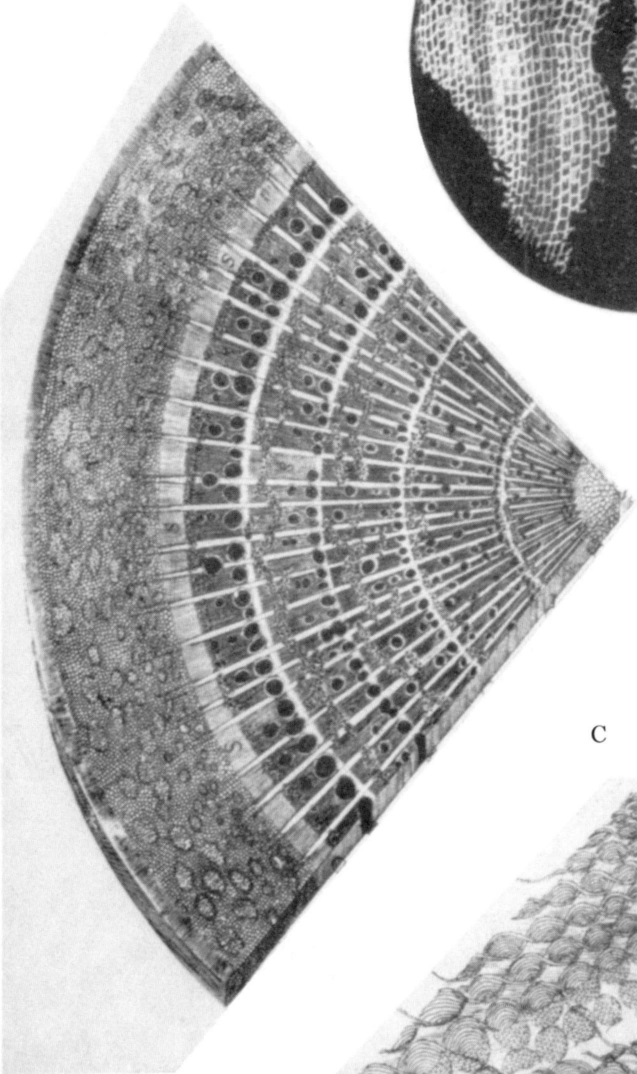

B

C

PLATE VIII

A

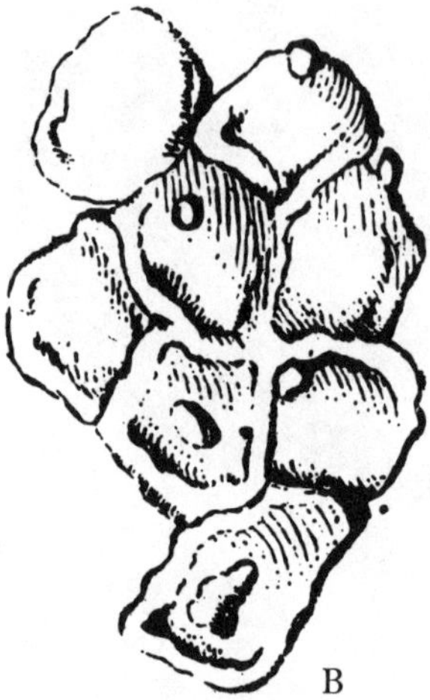

B

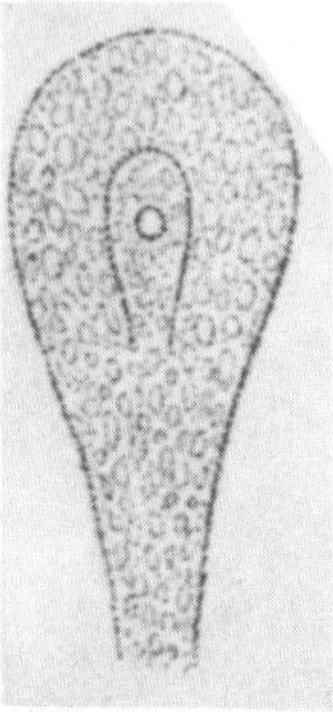

C

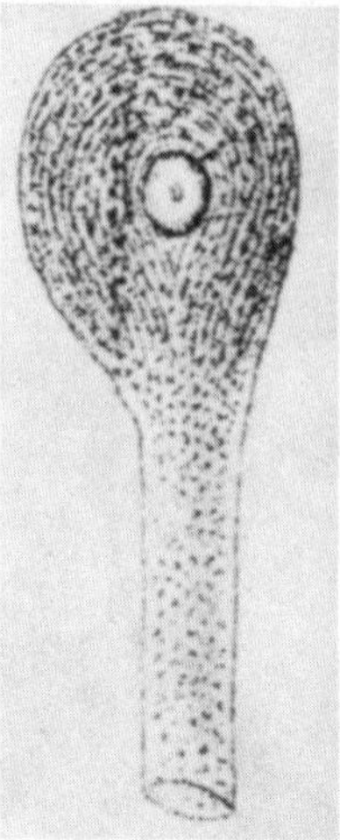

D

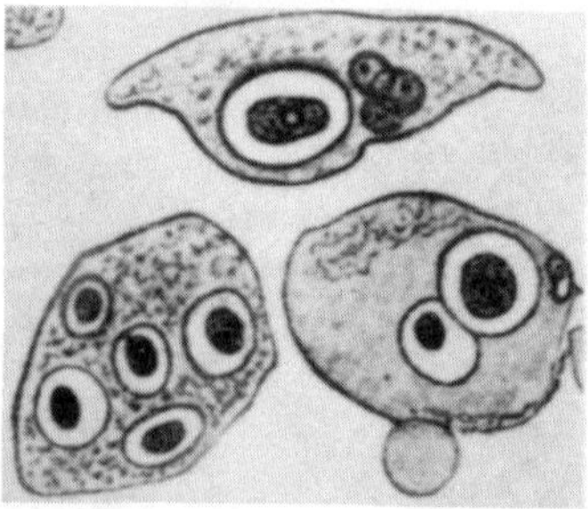

E

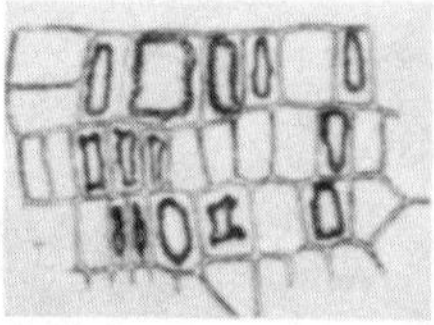

J

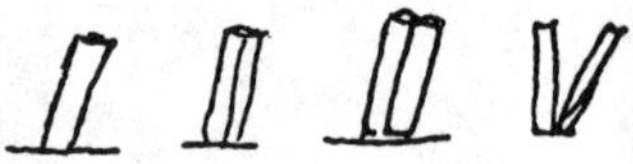

I

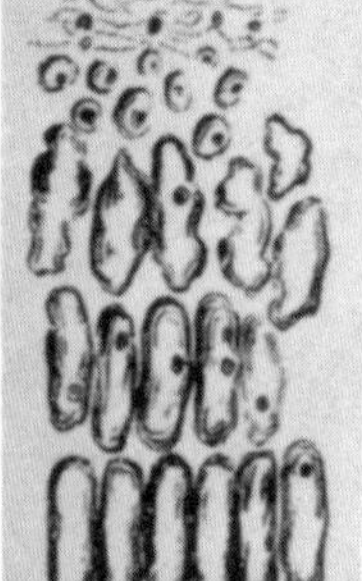

F

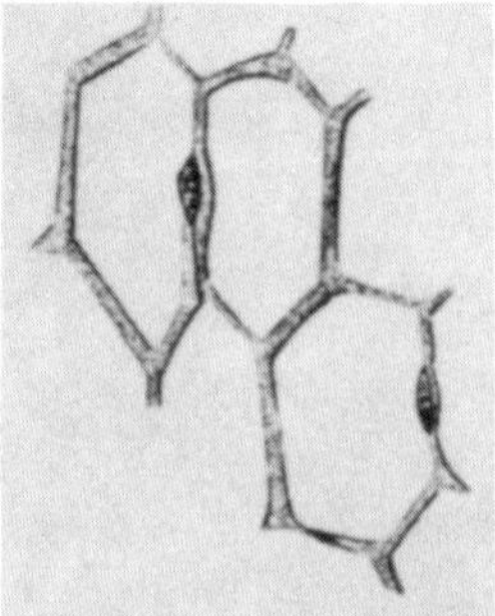

G

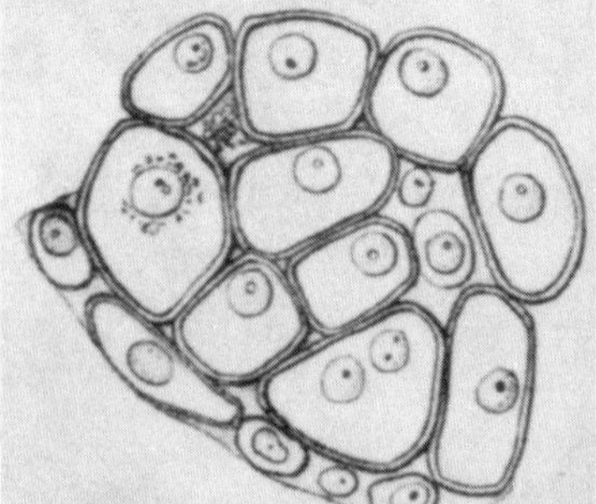

H

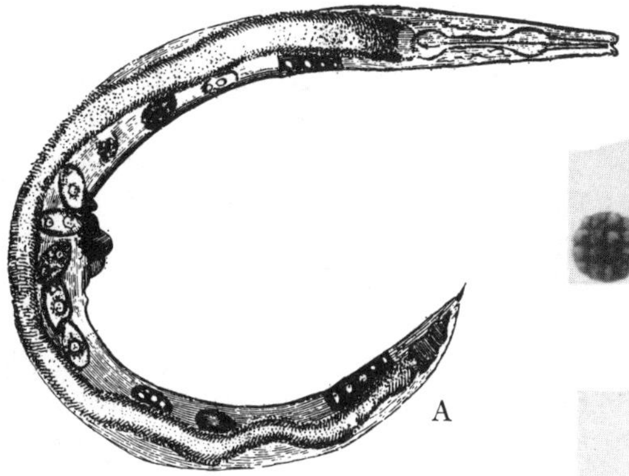

A

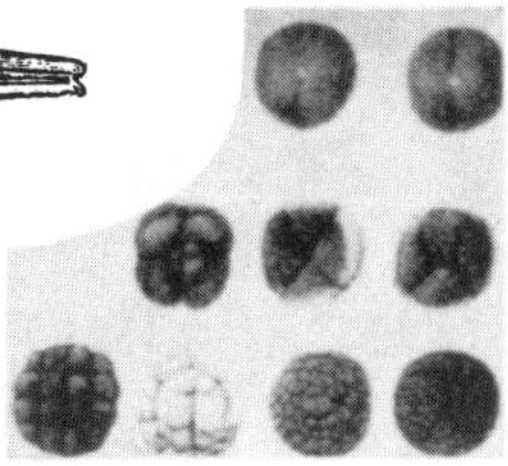

B

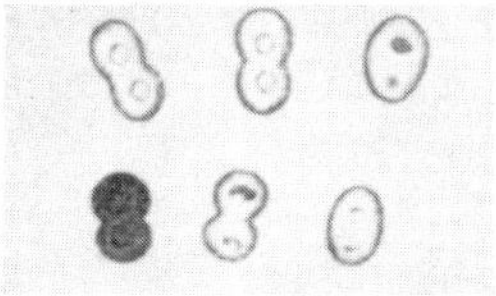

H

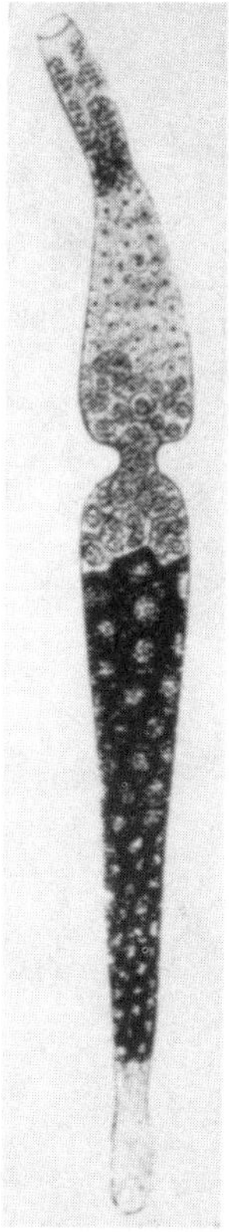

G

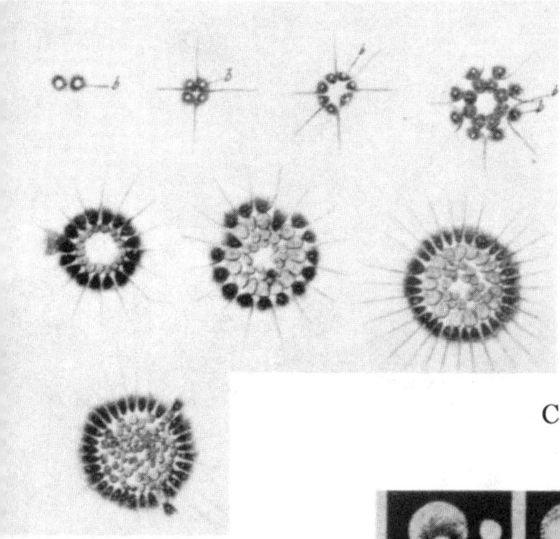

C

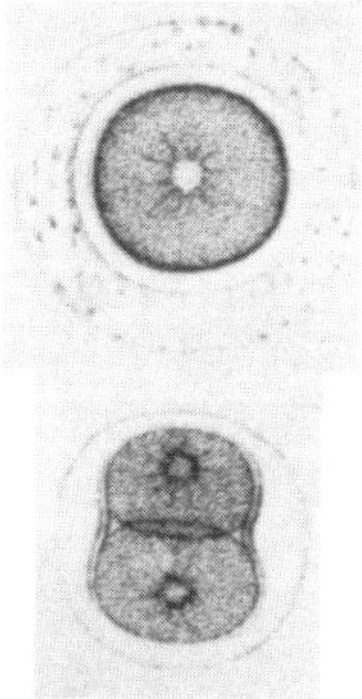

F

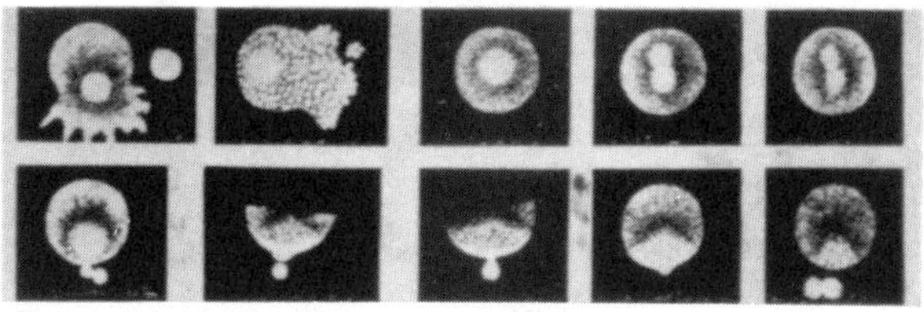

D

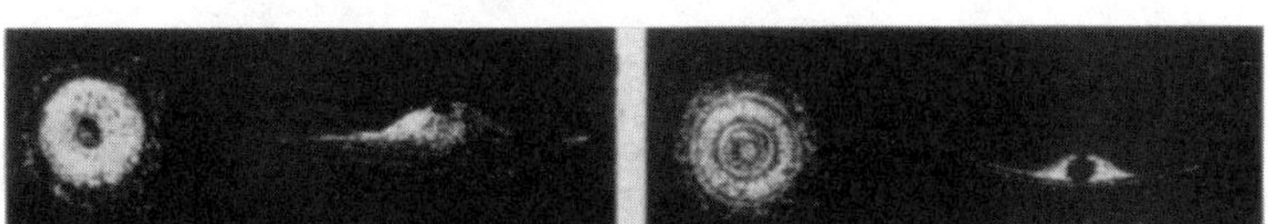

E

PLATE X

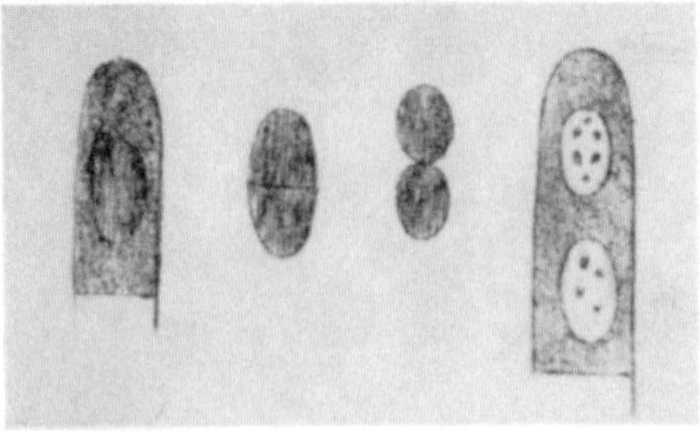

A

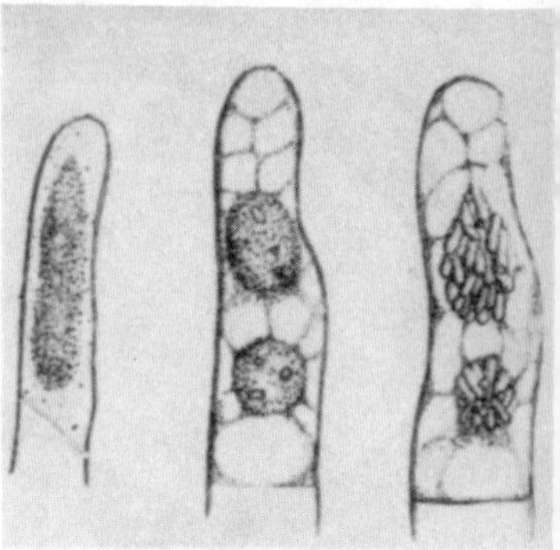

B

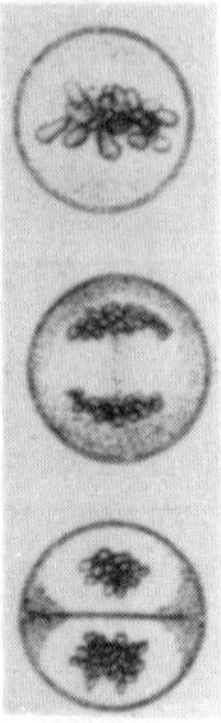

D

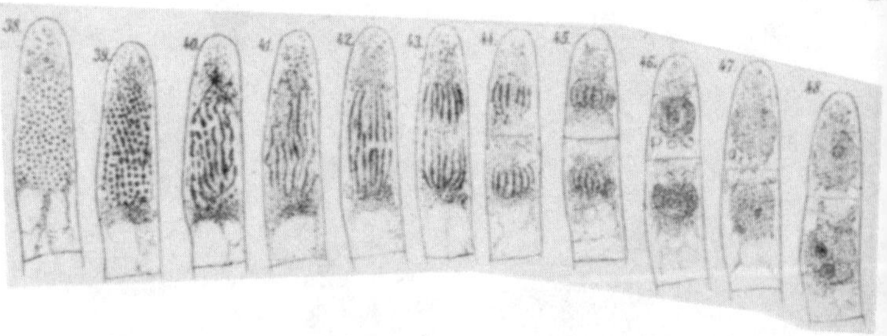

C

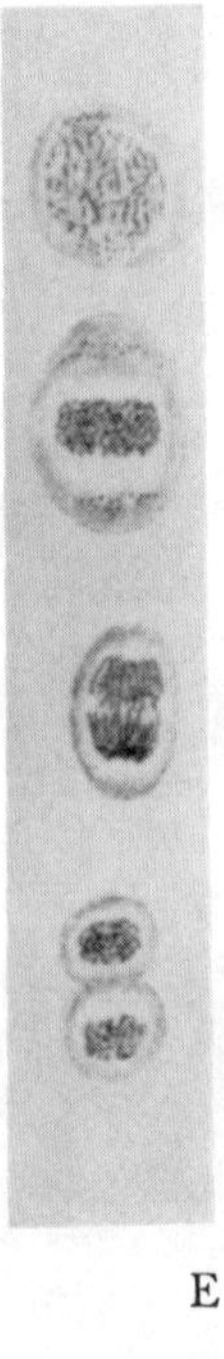

E

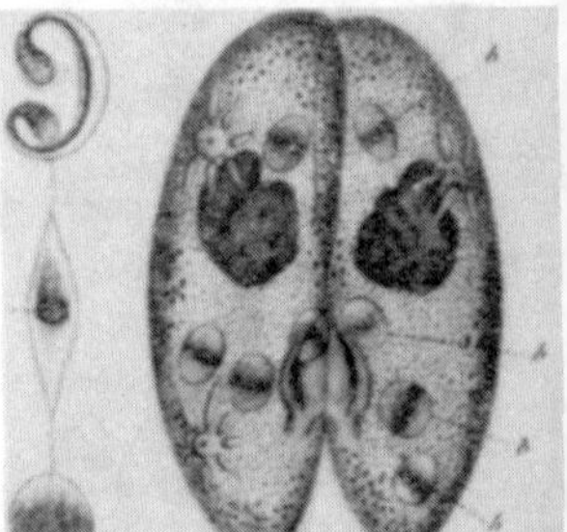

G

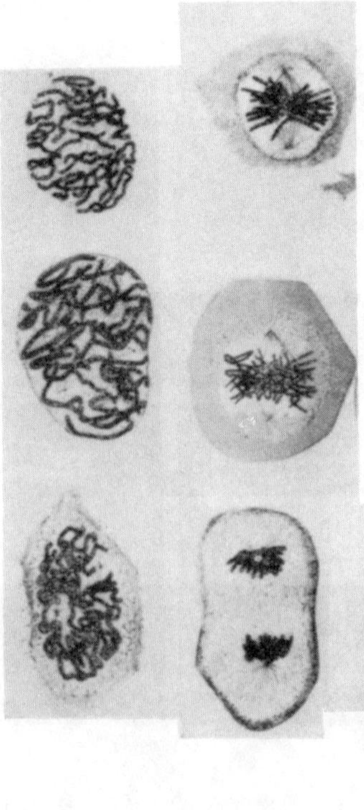

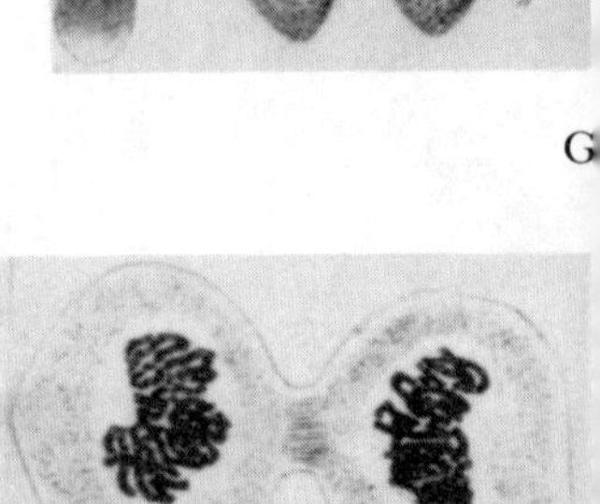

F

PLATE XI

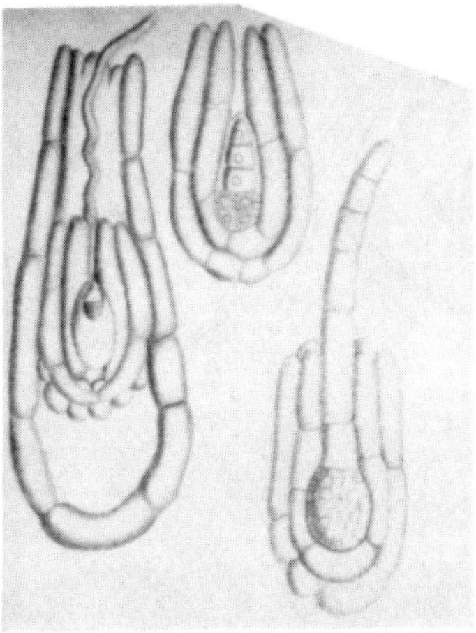

A

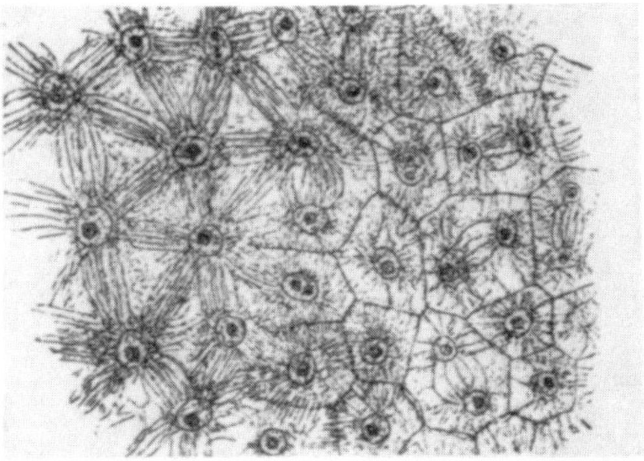

B

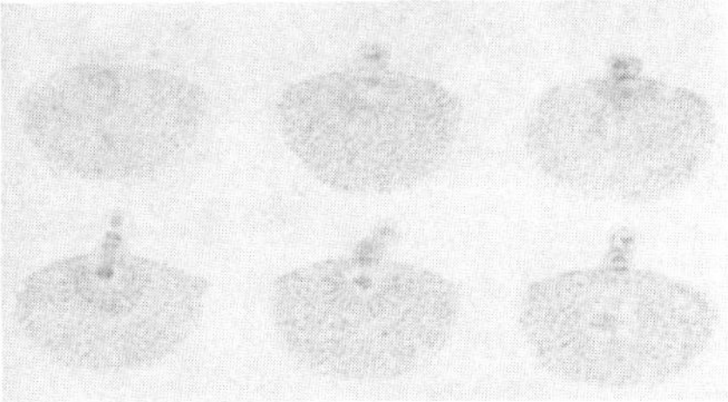

C

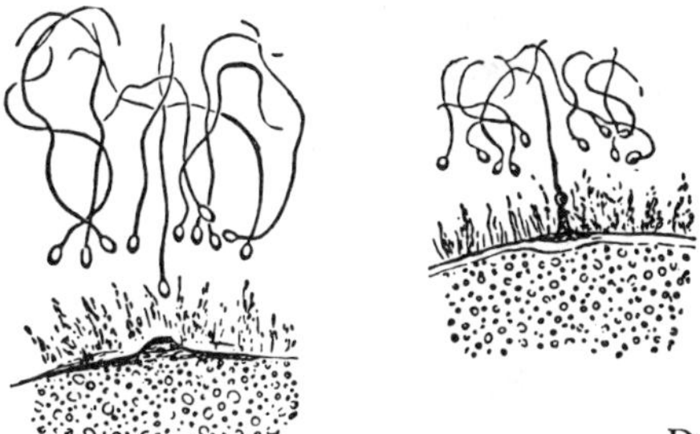

D

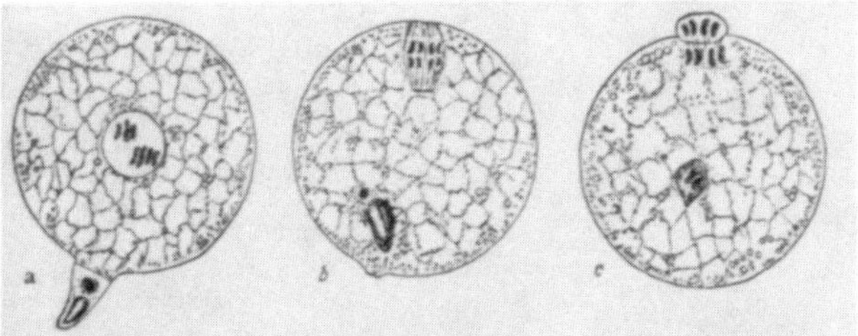

E

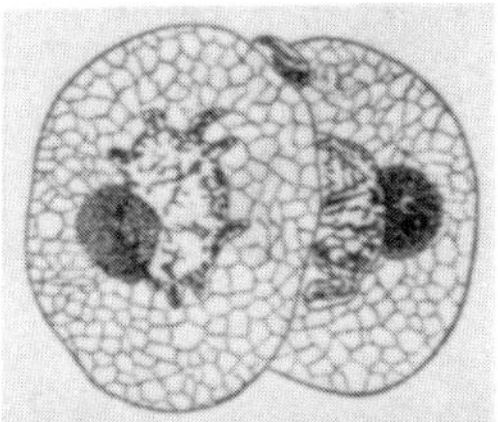

F

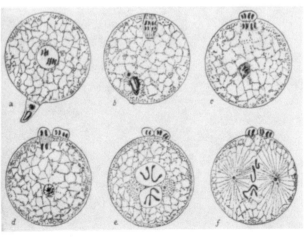

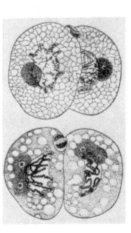

PLATE XII

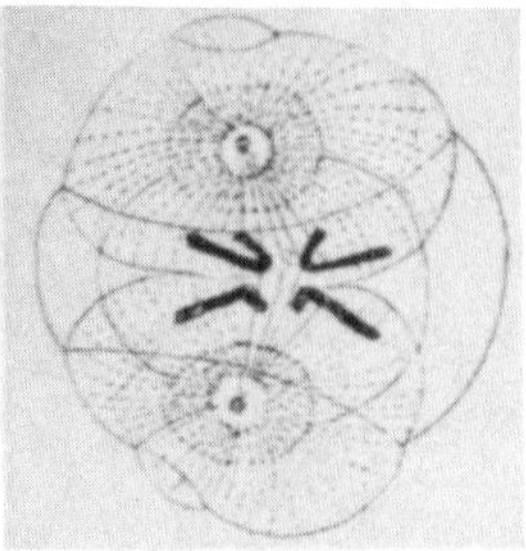

A

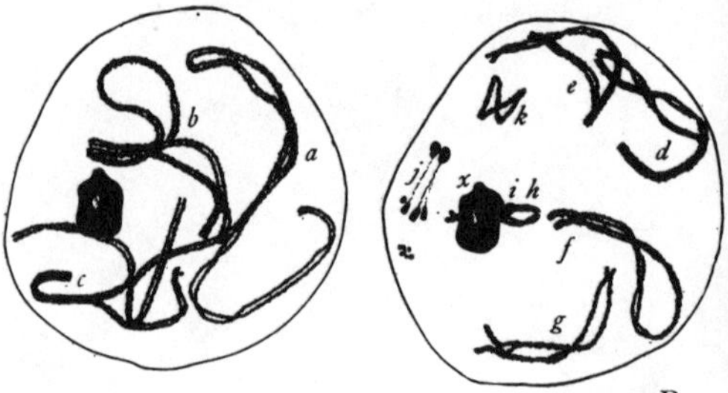

B

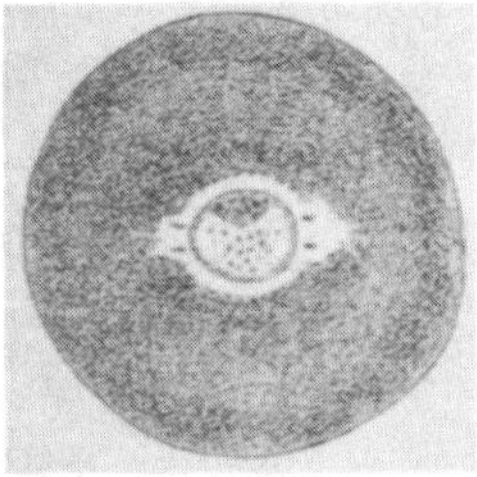

C

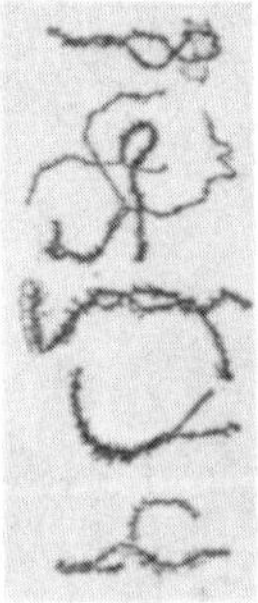

D

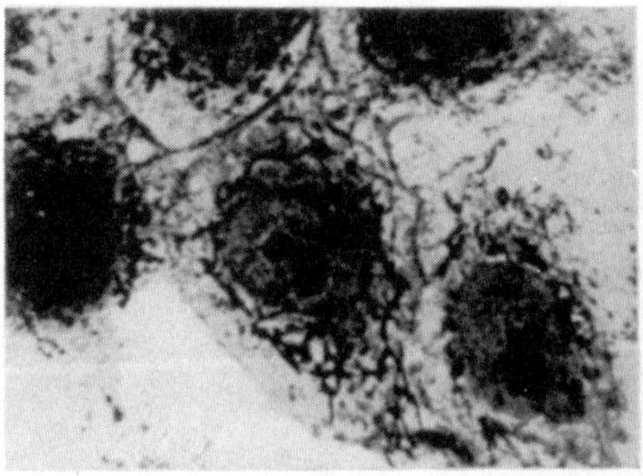

E

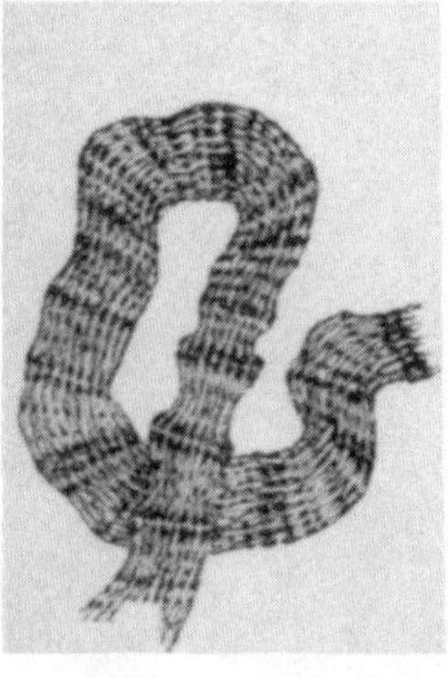

F

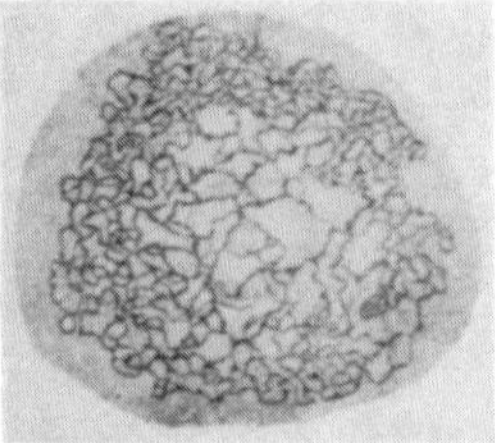

G

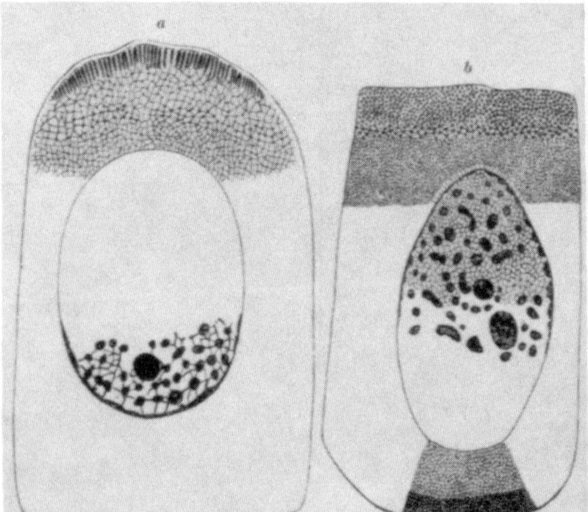

H

Index